# Applications of Photovoltaics

# Applications of Photovoltaics

Edited by R Hill
*Newcastle Polytechnic, Newcastle upon Tyne*

Adam Hilger, Bristol and Philadelphia

*British Library Cataloguing in Publication Data*

Applications of photovoltaics
   1. Solar energy. Photovoltaic conversion
   I. Hill, R. (Robert), *1937–*
   621.31′244

   ISBN 0-85274-277-0

*Library of Congress Cataloging-in-Publication Data are available*

Consultant Editor:
   **Dr R H Taylor**, Central Electricity Generating Board

Published under the Adam Hilger imprint by IOP Publishing Ltd
Techno House, Redcliffe Way, Bristol BS1 6NX, England
242 Cherry Street, Philadelphia, PA 19106, USA

Printed in Great Britain by J W Arrowsmith Ltd, Bristol

# Contents

| | | |
|---|---|---|
| List of Contributors | | vii |
| Preface | | ix |

**1 TERRESTRIAL APPLICATIONS OF PHOTOVOLTAICS** — **1**
*Michael R Starr*

| | | |
|---|---|---|
| 1.1 | Introduction | 1 |
| 1.2 | Photovoltaic Systems | 2 |
| 1.3 | Prospects for Photovoltaics | 13 |
| 1.4 | Obstacles to Progress | 22 |
| 1.5 | Conclusion | 24 |
| 1.6 | References | 25 |

**2 PHOTOVOLTAICS FOR DEVELOPMENT: IDENTIFYING REAL OPPORTUNITY** — **26**
*Phil O'Keefe*

| | | |
|---|---|---|
| 2.1 | Introduction | 26 |
| 2.2 | Third World Energy | 26 |
| 2.3 | Enduse Analysis | 27 |
| 2.4 | Towards a Structural Model for Photovoltaic Diffusion | 30 |
| 2.5 | Rapid Appraisal for Energy Projects | 32 |
| 2.6 | Social Evaluation of PV Systems | 34 |
| 2.7 | Conclusion | 37 |
| 2.8 | References | 38 |

**3 PHOTOVOLTAICS FOR DEVELOPING COUNTRIES** — **39**
*Bernard McNelis*

| | | |
|---|---|---|
| 3.1 | Introduction | 39 |
| 3.2 | Water Pumping | 40 |
| 3.3 | Photovoltaic Refrigerators for Rural Health Care | 50 |
| 3.4 | Lighting | 63 |
| 3.5 | Rural Electrification | 66 |
| 3.6 | Other Applications | 77 |
| 3.7 | Conclusions on Photovoltaics for Developing Countries— Summary of Experience | 83 |
| 3.8 | Conclusions | 86 |
| 3.9 | References | 87 |

**4 PHOTOVOLTAIC SYSTEMS FOR PROFESSIONAL APPLICATIONS** — **91**
*Alan Dichler*

| | | |
|---|---|---|
| 4.1 | Introduction | 91 |
| 4.2 | Photovoltaic Technology—Minimising Generator Costs | 91 |

| | | |
|---|---|---:|
| 4.3 | Cathodic Protection | 97 |
| 4.4 | Telecommunications | 99 |
| 4.5 | A Key to Rural Development | 101 |
| 4.6 | Hybrid Systems | 105 |
| 4.7 | A Practical Hybrid Application | 114 |
| 4.8 | Conclusions | 121 |
| 4.9 | References | 121 |

**5 LOW POWER APPLICATIONS OF PHOTOVOLTAICS**     **122**
*N M Pearsall and R Hill*

| | | |
|---|---|---:|
| 5.1 | Introduction | 122 |
| 5.2 | Conversion of Light to Electricity | 123 |
| 5.3 | Consumer Products | 128 |
| 5.4 | Replacement Power Supplies | 133 |
| 5.5 | Instrumentation | 141 |
| 5.6 | The Low Power Applications Market | 142 |
| 5.7 | Future Prospects | 144 |
| 5.8 | References | 144 |

**Index**     **147**

# List of Contributors

Alan Dichler | Solapak Ltd, Factory Three, Cock Lane, High Wycombe, Buckinghamshire, HP13 7DE

Robert Hill | Newcastle Photovoltaics Applications Centre, Newcastle upon Tyne Polytechnic, Newcastle upon Tyne, NE1 8ST

Bernard McNelis | IT Power Ltd, The Warren, Bramshill Road, Eversley, Hampshire

Phil O'Keefe | Department of the Environment, Newcastle Polytechnic, Ellison Place, Newcastle upon Tyne, NE1 8ST

Nicola M Pearsall | Newcastle Photovoltaics Applications Centre, Newcastle upon Tyne Polytechnic, Newcastle upon Tyne, NE1 8ST

Michael R Starr | 23 Gloucester Street, Faringdon, Oxfordshire, SN7 7JA

# Preface

The photovoltaics industry is growing rapidly throughout the
world. The previous markets in professional systems for
telecommunications, cathodic protection etc. are still
growing, whilst the applications in developing countries
have passed beyond the demonstration stage into an
acceptance of photovoltaic systems as a reliable engineering
solution to specific problems. In recent years, the most
rapidly growing market segment has been in consumer
products, but this is likely to be overtaken by the grid-
connected utility applications over the next few years.
Photovoltaics, in research, in production and in
applications has developed such momentum that its future is
assured as a major technology.

This book deals with the whole range of terrestrial
applications of photovoltaics. It is intended as an
introduction to the field for those with some technical
background but no specialist knowledge. As the range of
applications of photovoltaics expands over the coming years,
many engineers will need to gain an understanding of what
photovoltaics can do, without neccessarily needing to
understand the behaviour of solar cells. This book treats
the solar cells as black rectangles which convert light into
electricity and concentrates on the ways in which this
electrical power can serve human needs.

Photovoltaics has emerged from the dreams of its enthusiasts
to be a reliable commercial product in a remarkably short
period of time and it is instructive to review how this
transformation came about. Photovoltaics shares its birth
with photography in Bequerel's classic experiment of 1839.
The next hundred years saw a few notable pioneers developing
the selenium and copper oxide cells with minor applications
as light sensors. In 1954, both the silicon and the copper
sulphide cell were discovered, and attempts were made to use
silicon cells to power remote telephones. The space
programme of the 1960's gave tremendous impetus to the
development of solar cells, and brought the technology to a
point where it could be seriously considered for terrestrial
power supplies when the oil crisis of 1973 caused government
to consider ambient energy sources.

In 1974, the US government brought together its leading
experts to consider if, and how, photovoltaics could make a
major contribution to US electricity supply. It was clear at
that time that cost reductions by a factor of about 100,
improvements in working lifetimes of 10 times or more and
factors of 2 or more in efficiency would be required for PV
generated electricity to be cost-competitive with that from
conventional generating stations. To the great credit of
those involved, these factors were seen as a challenge and
not as reasons for discarding the technology. Programmes of

research, development and demonstration were set up and
received a massive boost under President Carter. The US
programmes were joined shortly afterwards by Project
Sunshine in Japan and then by the programmes of the European
Community and those of some individual European countries.

This injection of funds into photovoltaics R, D & D in the
late 1970's and early 1980's laid the foundation for the
present industry. The lessons learnt in module manufacture
increased the lifetime from months to years and the rigorous
quality control standards developed at JPL gave confidence
in the product to both manufacturers and purchasers. The new
approaches to cell production such as low cost silicon
feedstock, polycrystalline wafers, silicon ribbon and thin
film technologies, using amorphous silicon, copper indium
diselenide and cadmium telluride, all received their initial
impetus in this period.

As the political importance of energy costs waned in the
1980's, the government funding of photovoltaics also
decreased. The technology had, however, advanced to the
point where it was winning an increasing number of
supporters, both in the political field and, more
importantly at that time, amongst large companies, who could
take the long term view necessary to see the development
through to a successful and profitable conclusion.

Photovoltaic companies in the 1970's were mainly small and
were started and run by enthusiasts. Solarex was the world's
first PV company and the world's largest in the 1970's. When
the founder, Joe Lindmayer, sold a block of his shares to
Standard Oil, he became the world's first PV millionaire.
Other oil companies joined in the field and their financial
and technical muscle were crucial in bringing the industry
to its present position. The industry now is dominated by
large companies who initially took a cautious attitude to
the new technology, but who now have a long term commitment
and intend to be a part of what will clearly become a very
large worldwide business.

Photovoltaics has been technologically driven since the
discovery of the silicon cell. This was essential in the
early years, because the technical goals to be met were so
challenging. It is still the case that the major excitement
at international conferences comes from new achievements in
material or device development. This is both a sign of the
vigour of the photovoltaics community and of the relative
immaturity of the industry. The vigour is evident in the
rapid technical advances being made, with cell efficiencies
now exceeding 30% and at least four possible routes to the
reduction of module costs below the goal of US$1/peak watt.
Since 1974, module costs have decreased from about $70/peak
watt to about $2/peak watt for amorphous silicon modules and
about $4/peak watt for wafer silicon modules.        The

efficiency goal of 10% for modules is easily met by those using silicon wafers where efficiencies of 13-14% are common and 16% has been demonstrated. The efficiency of the amorphous silicon modules has yet to exceed 10%, being typically 6-7%, but efficiencies are improving continually. For the standard module with silicon wafers, lifetimes can be guaranteed to be over 10 years and studies of the mean time between failure suggest that modern module production techniques will result in lifetimes in excess of 20 years.

The lifetime and efficiency goals set in 1974 have been exceeded, whilst the cost goals are close. Furthermore, there is a long train of innovation, from possible world-beaters in the imagination of researchers to advanced working cells in the research laboratories, to pre-production modules in company laboratories, to new production facilities under construction, which secure a continuing advance of PV technology into the next century. However, the industry has now reached a stage where it must become market led and devote at least as much time on marketing and financing as it does to technology. There are markets for the present modules and systems at present prices which are very much larger than the present sales and it is the task of the industry, over the next few years, to increase the penetration into these markets as well as taking advantage of the new opportunities which will arise as a result of technical progress.

It is in this spirit that this book has been compiled. It deals only with the applications of photovoltaics, since there are already many excellent books on the theory and the technology. The book seeks to provide a comprehensive account of the terrestrial applications of photovoltaics, ranging from professional applications, developing country applications and consumer and leisure products.

The first chapter in this book gives an overview of the applications of photovoltaics for power supplies from 100 watts and upward. The third chapter concentrates specifically on applications in developing countries and explores in detail the uses where it is economically competitive with alternative power sources. Both of these chapters are written by distinguished engineers with many years of experience in the field and their chapters reflect their concerns to specify cost-effective services for the end-user. The fourth chapter is written by a leading systems engineer, whose concerns are to design and manufacture the PV systems and to identify markets for their products. This chapter is therefore orientated to the hardware required to provide a reliable and cost-effective service and this often involves the combination of photovoltaics with other power sources. Although there is necessarily some overlap between these chapters, the different viewpoints from which they are written provides an insight into the different concerns of the various sectors of the photovoltaics industry.

Photovoltaics is a vitally important energy source for the Third World, but it would be cruel to the two thirds of our fellow human beings who live in poverty if the technology were to be oversold. The second chapter explores the areas of Third World needs which can be addressed by photovoltaics and thus places the technology in context within the overall demand for energy services in these countries. It is as a tool for social and economic development rather than as a major energy source that photovoltaics can best play its part in rural areas of developing countries. However, it must be recognised that the "Third World" is far from homogeneous and countries such as India, Pakistan, Brazil or Algeria could well find cost-effective uses for bulk power from photovoltaics.

The final chapter discusses the application of photovoltaics in the consumer product and leisure industries. This has sometimes been dismissed in the past as the "gadget market", but it is a significant factor in the development of photovoltaics. It is at present the largest sector by value of the photovoltaics market and it has made a significant contribution to the cash flow of amorphous silicon manufacturers enabling them to fund further development. This market sector is also important in that it brings the general public into contact with photovoltaics in the context of a reliable and desirable product. It thus acts as a very good advertisement for photovoltaics as a whole, increasing general awareness and confidence in the technology.

In 1891, Rollo Appleyard was so enthused by solar cells developed by Sir George Minchin in London, that he wrote, in a letter to the editor of the 'Telegraphic Journal and Electrical Review', "Behold the blessed vision of the sun, no longer pouring his energies unrequited into space, but by means of photoelectric cells and thermopiles, these powers gathered into electric storehouses, to the total extinction of steam engines and the utter repression of smoke".

The environmental concerns of Appleyard were of a lesser magnitude than our present fears of acid rain, nuclear accidents and global warming, but his sentiments are as relevant today as they were nearly a hundred years ago. However, the stunning technological advances of the past fifteen years and the promise of the next fifteen give hope that his dream may be realised, and the application of photovoltaics will make a major contribution to the development of a sustainable future for mankind.

RH/November 1988

# 1

# Terrestrial Applications of Photovoltaics

Michael R. Starr

Consulting Engineer
23 Gloucester Street
Faringdon, Oxon, SN7 7JA, UK

## 1.1 Introduction

The photovoltaic effect, the direct conversion of light energy into electricity by solar cells, was first observed in 1839, but until the mid-1950's it remained largely a laboratory curiosity with relatively few practical uses. Some work was carried out by various laboratories to see if practical devices could be developed for battery charging, but the real breakthrough came with the space programme, starting with America's Vanguard I in 1958. Since then, practically all the many hundreds of scientific, commercial and military satellites launched by the various space organisations have been powered by silicon photovoltaic cells.

Following the 1973 oil crisis, interest in photovoltaics as a terrestrial source of power increased greatly and many countries, including several developing countries, instituted photovoltaic research, development and demonstration programmes. For example, in addition to the investments made by the manufacturers and other private interests, the US government increased its annual photovoltaic R & D budget from about $1 million in 1974 to more than $100 million per annum in the late 1970's and, despite recent cut-backs, still exceeds $30 million per annum. Although the level of funding has been somewhat less in Japan and in the European Community, there have been similar co-ordinated programmes of photovoltaic research and development. Total world expenditure from all sources on photovoltaic research, development and demonstration activities is currently (end 1987) estimated to be running at between $200 and $300 million per annum.

Photovoltaic arrays for space satellites are assembled to very exacting standards, as low weight and high reliability are of the utmost importance. Costs are consequently very high, with each Watt of power costing several hundred US dollars, but it has to be recognised that the total solar array cost for a satellite is relatively small in relation to the total project cost. The key issues for satellites are the power-to-weight ratio and reliability, not costs.

Much less expensive photovoltaic systems have been developed
for terrestrial applications, where the environmental and
other constraints are not nearly so onerous as in space.
Over the last ten years, there has been more than a tenfold
reduction in the real price of photovoltaic modules. This
has been achieved through a combination of improved cell
technologies and larger manufacturing volumes. Starting from
virtually zero in 1974, sales of photovoltaic systems have
grown to about 32 MWp in 1987, with a total value of at
least $500 million. Worldwide there are over 20 module
manufacturers of significance and there are several times
this number of firms designing and marketing photovoltaic
systems using bought-in components.

Costs have been steadily falling in real terms to the point
where photovoltaic systems are now becoming competitive with
other means of providing electrical power for an ever-
widening range of applications, from small calculators and
battery chargers, through navigation lights, water pumps and
cathodic protection systems, to generators for remote
buildings and island communities. New applications are being
found for which other forms of electricity generation are
quite unsuitable. There are good prospects for further cost
reductions, which will open up applications for
photovoltaics where the markets are very large indeed, such
as rural electrification and even, in time, grid generation.

## 1.2 Photovoltaic Systems

### 1.2.1 Market Categories

Photovoltaic systems provide a convenient and cost-effective
solution for the provision of relatively small amounts of
power needed for a wide range of applications. Needing no
fuel and very little maintenance and with no harmful
pollution at the place of use, such systems offer many
attractive features in comparison with the possible
alternatives. In some cases, photovoltaic systems provide
the only acceptable method of providing the necessary
electrical power at a remote site.

At present there are three market categories for
photovoltaic systems. First there is the large and growing
consumer market, for small electronic devices, garden lights
dry-cell battery chargers, and such items. Sales in this
market are largely dependent on good design, effective
marketing and reasonable prices. Secondly, there is the
market for professional systems, such as telecommunication
links, remote sensing, cathodic protection, navigation
lights, military equipment, etc. These systems normally
have to be justified on the basis of life-cycle costings
using conventional economic criteria, although environmental
considerations can often be an important factor. Thirdly,
there is the very large potential market for systems which

primarily have a social benefit, such as the provision of electricity for remote houses and villages, pumps for water supplies and irrigation, emergency telephone links, etc. Although these systems are expensive ($12-20 for every Watt of installed array power), the photovoltaic solution can provide important social benefits to a needy community, with lower life-cycle costs (and far lower operational problems) than diesel generators or grid-extension.

In the longer term, as system costs continue to fall in real terms, a fourth market category is expected to open up, namely that for grid-connected systems providing electrical power to buildings of all types or serving as central generators. Already a number of large commercial systems operating as intermediate peaking plant have been built in the USA to serve the grid, but the financial viability of these systems is currently strongly dependent on the level of tax credits and other inducements available.

## 1.2.2 Stand-alone Systems

There are currently many thousands of small stand-alone photovoltaic systems operating throughout the world for power (as distinct from consumer) applications. They range in size from a few Watts to several tens of kilowatts, plus many millions of solar powered calculators and other consumer products. Power for remote radio, television and microwave repeater stations may in many cases be economically provided by photovoltaic generators. Various experimental and demonstration plants have been built and now this application is considered by the operating companies to be the appropriate solution on technical and economic grounds for many cases (Fig. 1.1). Several long stretches of desert roads in the Middle East have been equipped with photovoltaic powered emergency radio-telephones at regular intervals. There are also numerous military applications for photovoltaics for telecommunications and remote sensing. Cathodic protection of pipelines crossing the deserts of the Middle East and elsewhere is another professional application which has proved commercially attractive in recent years (Fig. 1.2).

Most photovoltaic manufacturers now offer a wide range of standard systems, for battery charging, water pumping, street lighting, domestic lighting, refrigeration, electric fencing, alarm and security equipment, remote monitoring beacons and other navigational aids; the list is constantly growing as other applications are being found. Although some further improvement and demonstration of these systems is continuing , in most cases they can be considered as developed products. Commercial sales are growing steadily to private and public customers, who find that photovoltaic systems provide the most economic or conventional solution to their needs.

Figure 1.1          Photovoltaic power for telecommunications
                    (source: BP Solar)

Figure 1.2          Photovoltaic power for cathodic protection
                    (source: BP Solar)

In addition to these relatively small standard systems, there are also many specially engineered larger systems, either completely stand-alone or working in conjunction with diesel generators or the grid. Many of these are pilot or demonstration plants, sponsored under various research programmes or as part of aid to developing countries. Most of the 15 EEC-sponsored pilot plants listed in Table 1.1 come into this category, such as the 100kWp system, working in conjunction with diesel and wind generators on the Greek island of Kythnos (Fig. 1.3).

The experimental or pilot phase has now moved onto the demonstration phase. Over the period 1983 to 1986, some 48 photovoltaic demonstration projects have now been built or are under construction in Europe with support from the EEC, with total array power about 720kWp. A number of these projects involve multiple installations, such as one scheme in France to bring electricity to 40 remote houses. These projects, which are listed in Table 1.2, cover a wide range of applications. All these demonstration projects, and additional projects being planned for the future, are intended to establish the technical and economic basis for similar systems to be installed in the future on a commercial scale.

The first solar powered television set for a village school was installed as long ago as 1968 in Niger. Since then several hundred schools in Niger have been similarly equipped and a number of other, mainly West African, countries have also initiated projects for solar powered educational television. The photovoltaic systems have generally proved more reliable and less trouble to operate and maintain than the primary batteries or petrol generators previously used. The social benefits are clearly very important.

Small-scale solar powered pumping systems for water supply and irrigation applications have long been considered of great importance for improving living conditions and raising agricultural output in developing countries (Fig. 1.4). A good market for solar pumps to serve livestock watering points has opened up in the southern states of the USA and in other countries where the cost of regularly servicing engine powered pumps is prohibitive. A number of relief agencies have also found photovoltaic pumps to be a convenient and on the whole reliable solution to the problems of water supply in remote refugee camps.

In view of the great interest in water pumping, the UNDP and World bank have sponsored a special project to test and demonstrate small-scale solar pumps of all types. After a four-year programme of laboratory and field testing, plus extensive economic and technical studies, clear guidelines

Table 1.1          Photovoltaic pilot plants supported by the
                   EEC (1981-1984)

| Project | Power | Site | Application |
|---------|-------|------|-------------|
| 1 | 50kWp | Aghia Roumeli Crete, Greece | Electrification of remote village |
| 2 | 63kWp | Chevetogne Belgium | Power for solar heated public swimming pool |
| 3 | 50kWp | Fota Island Ireland | Power for dairy farm, in conjunction with grid |
| 4 | 45kWp | Giglio Island Italy | Water disinfection and cold-store for agricultural produce |
| 5 | 30kWp | Hoboken Belgium | Hydrogen production and water pumping in industry |
| 6 | 35kWp | Kaw French Guyana | Electrification of remote village |
| 7 | 100kWp | Kythnos Island Greece | Power supply to island network, in conjunction with wind and diesel |
| 8 | 30kWp | Marchwood Great Britain | Power supply to grid |
| 9 | 50kWp | Mont Bouquet France | Power supply to TV transmitter |
| 10 | 50kWp | Nice France | Power for Nice airport technical systems |
| 11 | 300kWp | Pellworm Island F R Germany | Power supply for holiday centre, in conjunction with diesels |
| 12 | 44kWp | Rondulinu Corsica, France | Electrification of remote village |
| 13 | 50kWp | Terschelling Island Netherlands | Power supply to marine training school, in conjunction with grid |
| 14 | 65kWp | Tremiti Islands Italy | Power supply for water desalination plant |
| 15 | 80kWp | Vulcano Island Italy | Power supply to island community |

Table 1.2    Photovoltaic demonstration plants supported
by the EEC

| Project No. SE/ | Country | Application | Location | Array kWp |
|---|---|---|---|---|
| **PV DEMONSTRATION PROJECTS 1983** | | 9 Projects, total array power 147.7kWp | | |
| 123 | IT | Fire prevention and high water alarm | L'Aquila & Pescara | 0.4 |
| 134 | DE | PV system in connection with small power heat cogeneration for an isolated farmhouse | Baden-Wurttemberg | 5 |
| 215 | HE | PV pumping for irrigation | Menetes Carpathos Island | 10 |
| 313 | HE | PV station for two remote villages | Gavdos Island | 20 |
| 314 | HE | PV station for groups of isolated houses | Antikythira Island | 35 |
| 404 | HE | PV station for a remote village | Arki Island | 25 |
| 466 | FR | PV rural electrification of 40 isolated houses | Southern France | 32 |
| 560 | DE | PV electricity supply for ground water level measurement | Hamburg | 0.3 |
| 704 | HE | PV electricity supply for an isolated relay station | Antikythira Island | 20 |
| **PV DEMONSTRATION PROJECTS 1984** | | 12 Projects, total array power 96.5kWp | | |
| 084 | DE/FR | PV electricity for three stand-alone residences | Baden-Wurttemburg, Alsace | 2.6 |
| 267 | DE | PV electricity for a bird and weather station on a small North Sea island | Scharhoern | 4.1 |
| 507 | UK | A PV solar/wind cogeneration system to contribute to domestic electricity supplies in mains connected houses | Milton Keynes | 4.2 |
| 520 | IR | Integration of a PV generator in mains connected residential systems | Farnanes, Cork | 5 |
| 555 | DK | PV solar/wind cogeneration plant for a remote community | Hasle, Bornholme Island | 36 |
| 573 | IT | PV and micro-hydro power study of an isolated farm | Campoligure, Ligur. Appennines | 2.1 |
| 624 | HE | PV and wind power supply of unmanned lighthouse | Lithari, Skyros Island | 2.6 |
| 648 | HE | PV and wind power supply of unmanned lighthouse | Methoni, Sapienza Pel. | 2.6 |
| 674 | IT | PV electricity supply of a lighthouse | Palmaiola Island | 3.9 |
| 692 | IT | PV electricity for Etruscan archeological sites | Cetano and Sorano | 26.1 |
| 696 | FR | PV electricity supply for alpine lodgings and huts | French Alps | 5.3 |
| 713 | DE | Stand-alone PV and wind electricity supply for various purposes in recreational areas | Freiburg im Breisgau | 2.1 |
| **PV DEMONSTRATION PROJECTS 1985** | | 14 Projects, total array power 170.6kWp | | |
| 098 | FR | PV electrification of a nature reserve | Camargue | 12 |
| 114 | FR | PV/diesel power supply to an isolated microwave relay | Toulon | 1.92 |
| 122 | FR | PV electrification of a TV and FM emitter | French Guyana | 20 |
| 138 | IT | PV electrification of 30 isolated houses | Eolian Islands | 9 |
| 143 | HE | PV power supply for desalination, refrigeration and lighting | Milos Island | 31.9 |
| 146 | IT | PV electrification of dairy farm in the mountains | Abruzzo | 6.3 |
| 159 | IT | PV powered fog detecting system for high risk highway | Verona | 14.3 |
| 187 | HE | Autonomous PV power supply for housing complex | Poros Island | 4.15 |
| 280 | DE | Grid-connected low power system | Saarbrucken | 8 |
| 412 | DE | PV electricity house supply for decentralized relief of mains supply | Berlin | 5.1 |
| 501 | FR | Solar PV and thermal energy for 29 high-mountain refuges | Alps, Pyrenees | 8.35 |
| 540 | IT | PV electrification of 7 dairy farms in high mountains | Liguria | 27.4 |
| 565 | IT | PV plant for the energy supply of a lighthouse | Sicilia | 18.2 |
| 593 | FR | Photovoltaic water pumping in rural area | Corsica | 3.84 |
| **PV DEMONSTRATION PROJECTS 1986** | | 13 Projects, total array power 307kWp | | |
| 103 | FR | PV electrification of isolated houses, TV repeater, harbour beacon, water steriliser | Corsica | 9.76 |
| 159 | FR | PV power supply for three telephone exchanges | Guadaloupe | 12.4 |
| 215 | PO | PV electrification of isolated rural dwellings | Portugal | 35 |
| 233 | ES | PV powered reverse osmosis desalination | Isla del Moro | 23.12 |
| 246 | FR | PV electrification of light buoys at different latitudes | Atlantic Ocean, Mediterranean Sea, Antilles | 24 |
| 261 | ES | PV power plant | Tabarca Island | 100 |
| 327 | ES | PV rural electrification of 57 houses | Sierra de Segura | 28.2 |
| 333 | ES | PV power for a cheese cooperative | Trebujena | 19.7 |
| 337 | ES | PV rural electrification of 30 houses | Ivares, Oden Lleida | 18.9 |
| 339 | ES | PV and wind powered reverse osmosis seawater desalination plant | Fuerteventura Island | 20 |
| 388 | IT | PV powered airport signalling lights | Lucca - Tassignano | 4.75 |
| 390 | IT | Passive thermo/PV hybrid system for climatic control and electrification of isolated buildings | Palombara Sabina | 3.0 |
| 493 | FR | PV equipment programme of the International Activity Park | Valbonne - Sophia Antipolis | 8.3 |

Figure 1.3      100kWp photovoltaic generator on Kythnos Island (source: CEC)

Figure 1.4      Photovoltaic water pump on test in the Philippines (source: Halcrow/IT Power)

have now been established to help potential purchasers
decide whether solar pumping is appropriate for their
circumstances and, if so, how they should set about
procuring suitable equipment (Ref. 1.1). In addition to the
technical and economic issues, there are also important
social implications to be considered, a subject examined in
one of the follow-up studies to the World Bank/UNDP project
(Ref. 1.2).

As a development of straightforward photovoltaic pumping
systems, complete water treatment systems incorporating
filtration and sterilization are now becoming available that
are self-contained and simple to operate. For small islands
or places where the water is brackish, a number of
photovoltaic-powered reverse osmosis desalination units have
been demonstrated and in certain circumstances this
technology could soon prove to be the most economic option,
taking into account the high cost and practical problems
involved with alternatives such as diesel generation or
water brought in by tanker trucks or boats.

Photovoltaic refrigerators have been under development for
many years and several types are now available commercially.
The World Health Organisation has been encouraging the
development of medical refrigerators suitable for storing
vaccines and other medical supplies, the principal technical
requirements being as set out in Table 1.3. Costs are still
too high for major sales but technical work is continuing to
improve performance and eliminate, if possible, the need for
batteries, always a weak point in a system which otherwise
needs little or no maintenance.

It is important to stress that a stand-alone photovoltaic
generator for a given application should not be considered
merely as a substitute for a battery or diesel generator.
All components interact and the objective of good system
design must be to obtain the most cost-effective and
reliable combination of components, taking into account the
expected variation in incident solar radiation and the
requirements of the end user. For some applications, the
additional cost and complication of battery storage may be
avoided by storing the end product, such as water or ice.
Often it will be found that pumps and motors of higher
efficiency than normally available will be more cost
effective, since the size of the expensive photovoltaic
array (and battery) may then be reduced. Similarly the
thickness of the insulation for refrigerators needs to be
optimised as part of the system design.

### 1.2.3 Rural Electrification

A large proportion of the rural population in most
developing countries does not have a proper electricity
supply. Even in Europe, where grid coverage is generally

Table 1.3        WHO specification for photovoltaic
                 refrigerators

Net vaccine capacity:      :    30/40 litres (top opening)

Ice making performance     :    Minimum 1kg/24hr in +32$^0$C ambient temperature

Refrigerating performance :     No part of the vaccine storage area to be
                                above +8$^0$C or below -3$^0$C in:
                                a) +43$^0$C ambient temp.
                                b) +32$^0$C ambient temp.
                                c) +43$^0$C day time and +15$^0$C night time cycle

Hold-over time             :    More than 6 hours below +10$^0$C when power cut
                                out in +43$^0$C ambient temp.

External casing            :    Non-corrodable

Minimum battery
maintenance interval       :    One year

Insulation                 :    Rigid polyurethane

1984 cost target           :    Less than US$1500 complete

considered to be complete, there are still many thousands of permanently occupied houses in all the Mediterranean countries which are not connected to the grid, plus many more holiday homes and tourist facilities. There is thus considerable scope for photovoltaic systems to provide power for basic household requirements in these areas, but at present the high capital cost of such systems is a formidable obstacle, particularly in view of the generally low income level of the households concerned. Although it may be assumed that rural customers in European countries could afford to pay rather more for electricity than their counterparts in developing countries, it would still be necessary for the photovoltaic generators to be subsidised to a considerable extent.

Such subsidies would not be unreasonable, in view of the benefits that would accrue. Bringing electricity to remote areas has long been viewed as a development priority in both industrialised and developing countries, leading as it does to improved living standards and the opening up of new prospects for industry and agriculture. Because of the high cost of extending the grid or installing local networks based on diesel generators, it is accepted that subsidies are needed to offset the high capital and running costs of rural electrification, so that consumers in remote areas end up paying for their electricity at rates similar to those obtaining in the urban areas.

For providing electricity to remote villages for which grid extension would be too expensive, two approaches are possible: either the whole village may be served from a central generator, or each house may be equipped with its own self-contained system. Most demonstration plants have taken the former approach, no doubt mainly to simplify administration and raise the publicity value, but there are considerable advantages in adopting the latter approach, as discussed by Starr (Refs. 1.3, 1.4), since the cost and complications of a distribution system may be avoided and the long term administrative overheads reduced. A 500Wp system would be sufficient in many developing countries to power the basic requirements of a typical household, with, for example, three lights, two fans, a television set and a small refrigerator. A 350Wp system would be sufficient if the refrigerator were omitted. Battery storage would be sized to suit the climate and degree of reliability required.

Such a system could be entirely direct current, at 24 or 48V, and thus avoid the need for an inverter, but this would create considerable problems for the users since they would not be able to use standard low cost AC domestic appliances bought in the local markets. The best solution may be to have a DC circuit for lights and an AC circuit for appliances, with the inverter automatically switched off to reduce losses when not required.

Stand-alone photovoltaic generators, complete with battery storage sufficient for three to four days supply, currently cost about $15-20/Wp if installed in large quantities. The levelized electricity unit cost works out to be about $0.90 to $1.20 per kilowatt-hour, depending on the interest rate. This unit cost is about ten times higher than the cost of grid-generated electricity in urban areas, but it is often not much higher than the true cost of running diesel generators or extending the grid to serve remote areas. Given continued progress with the development of low cost photovoltaic cells and, equally important, substantial cost reduction for batteries and other balance-of-system components. It should be possible within 5 to 10 years to bring costs down to $7/Wp or less. At this price, and assuming a level of subsidy comparable to that given directly or indirectly for conventional systems, photovoltaics would begin to be a viable proposition for large-scale rural electrification. The market potential for this application in developing countries is enormous.

In industrialised countries, stand-alone systems for off-grid houses usually need to be of considerably higher rating than those in developing countries, to allow for much higher energy use. Where necessary, air-conditioning imposes a particularly heavy load and allowance may also need to be made for appliances such as freezers, washing machines, irons and hair dryers. Ideally, the design of the whole house needs to be optimised, incorporating passive features and energy efficient appliances. A number of such houses, some private and some publicly-funded as demonstrations, have been built in the USA and in European countries, with passive features and solar thermal and photovoltaic systems.

Electricity unit costs somewhat higher than in developing countries would probably be acceptable for providing electricity to remote homes in Europe in places where grid extension is not feasible. Higher prices would also be acceptable for holiday homes and for other buildings serving the tourist industry.

In the short term, whilst system costs remain too high for large-scale rural electrification schemes, it would be best to concentrate available funds on selected applications for photovoltaics which offer good demonstration opportunities coupled with high social benefits. Small systems capable of powering only one or two lights are often found to be popular in areas at present relying on kerosene and they can significantly improve the quality of life. Photovoltaic-powered telephones can link remote villages with the trading and administrative centres and thus bring a wide range of benefits. Other high-value applications would be power for water supplies and health centres. A good example of the latter is at San in Mali, where a 9.8kWp photovoltaic generator was installed in 1979 to provide all the electricity needed by a 100-bed hospital, including power for the water supply system (Fig. 1.5).

### 1.2.4 Grid-connected Systems

There has been much speculation on the issue of whether photovoltaics will eventually become cheap enough to be economic for grid-connected applications. At present, with oil and coal prices depressed, this seems to be a remote prospect, but in the long term the position could change, particularly for countries rich in solar energy but low in conventional fuels and unwilling (or unable) to introduce nuclear technology. Many countries have made a strong political commitment to encourage the use of renewable energy resources and some have gone further by deciding not to build any new nuclear power plants (eg, Sweden).

Grid-connected systems are simpler and less expensive than stand-alone systems, since they require little or no battery storage. The grid itself can serve as "storage", with the photovoltaic plant supplying power to or drawing power from the grid depending on the load and solar irradiance.

There are a number of grid-connected photovoltaic systems in the USA, including several private residences, some public buildings and a few multi-megawatt generating plants in California (Figs. 1.6 and 1.7). Some of these plants were built for demonstration purposes, but others could be justified financially because of the tax credits and other incentives available from the federal and state authorities. An important factor influencing the economics of the central generators was the relatively high price the utilities were prepared to pay for the electricity from the solar (and wind) generators, faced as they were with the urgent need to install additional capacity to meet peak demand. In some cases, these problems had arisen as a consequence of the delays in the construction programme for large fossil fuel or nuclear plants. As the system of tax credits is being phased out, photovoltaic generators will not be financially viable for grid-connected applications, at least until system costs come down from the present $7 to 10/Wp to around $2 to 3/Wp.

### 1.3 Prospects for Photovoltaics

### 1.3.1 Photovoltaic Sales

It is not easy to obtain reliable data from manufacturers on their current sales of photovoltaic modules and systems. However, after reviewing various published estimates (eg, Ref. 1.5) and from discussions within the industry, it seems

Figure 1.5      9.8kWp   photovoltaic   generator   for   San
                hospital, Mali (source: IT Power)

Figure 1.6      6.5MWp PV central generating plant, Carrisa
                Plains, California (source: ARCO Solar)

clear that about 32MWp of photovoltaic systems were sold in 1987, worth at least US$500 million. The growth in sales over the last ten years is illustrated in Fig. 1.8. The very high growth rate in the 1970's dropped to about 30 percent per annum in the early 1980's and is now around 10 percent per annum.

An approximate breakdown of the total market is set out in Table 1.4. The largest single item is accounted for by the small consumer products. These are mostly powered by amorphous silicon cells. Sales of solar powered calculators, transistor radios, cassette players, clocks and other electronic items are growing steadily and new consumer products are constantly emerging. A particularly successful item that came on the market in 1987 is the individual garden light, powered by a 2 or 3Wp amorphous silicon cell. It is estimated that, in terms of power, amorphous silicon cells comprised about one-third of the total world market in 1987.

The market for photovoltaic generators for remote telecommunications installations was the second largest sector in 1987, at about 10MWp. Continued growth is expected in this area, as photovoltaic systems are ideal for providing the relatively small amounts of power required in remote locations. These systems are almost all built using crystalline silicon cells, since these are well-proven and cost-competitive with amorphous silicon cells on power applications.

## 1.3.2 Technology and Prices

Over the last ten years, the industry has achieved more than a tenfold reduction in the price per peak Watt of photovoltaic modules. The dominant cell technology has been crystalline silicon, initially in mono-crystalline form but in recent years several manufacturers have been moving over to semi-crystalline cells. The price reductions have been achieved by a combination of improved cell technologies and larger production volumes. There has not been any significant decrease in the cost of the electronic grade polysilicon feedstock, as attempts to develop cheaper solar grade silicon have been largely unsuccessful to date.

Module prices for both forms of crystalline silicon are currently around $4.5 - 5.5/Wp for large orders, ex works. Bearing in mind that the cells account for about 60 per cent of the module price, some further price reductions, possibly down to $3/Wp or less, are foreseen through the introduction of cheaper silicon and larger, fully automated, manufacturing plants. Some manufacturers will continue to favour the well established mono-crystalline technology, others will move over to the semi-crystalline approach, particularly if solar grade silicon becomes available. It

Figure 1.7        300kWp  grid-connected  system,  Georgetown
                  University, Washington DC, USA (source: M.R.
                  Starr)

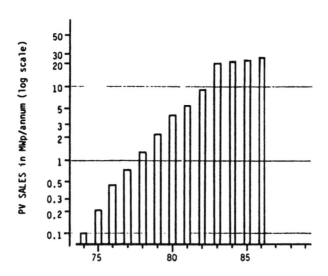

Figure 1.8        Photovoltaic sales 1974-1987

Table 1.4          **Estimated sales of photovoltaics in** 1987

| Application | Sales in MWp |
|---|---|
| Telecommunications | 9.0 |
| Cathodic protection | 1.0 |
| PV/Diesel hybrid generators | 4.0 |
| Water pumps | 0.5 |
| Navigational aids | 1.5 |
| Off-grid residential - USA | 0.5 |
| Grid-connected residential - USA | 0.1 |
| Rural electrification (excl USA) | 3.0 |
| Refrigerators | 0.2 |
| Military applications | 0.2 |
| Central generators - USA | 0.5 |
| Consumer products - crystalline Si | 0.5 |
| Consumer products - amorphous Si | 11.0 |
| TOTAL ESTIMATED SALES IN 1987 | 32.0 MWp |

is noteworthy that the 9MWp/annum plant opened by Hoxan in Japan in 1984 uses mono-crystalline technology. At present only one manufacturer, Mobil Solar Energy Corporation, USA, has a commercially crystalline sheet process, but others are continuing development work and one of these could yet emerge as being competitive with the ingot processes.

Much lower costs, even down to $1/Wp or less, are potentially attainable with thin film cells. In view of the large efforts being made world-wide to develop various thin-film technologies, it is probable that sooner or later large area thin film cells will become available with much improved efficiency and stability than current products. At present, amorphous silicon dominates the thin film scene, with commercial module efficiencies typically in the range 4 to 6 per cent, considerable less than the 10 percent or more available with crystalline silicon technology. There is also the problem of light-induced degradation, which in practice means that the initial module power has to be derated by about 20 percent for design purposes. The ex-works price of amorphous silicon modules suitable for power applications (as distinct from consumer products) is only a little less than the corresponding price for crystalline silicon modules. This means that the installed cost of amorphous silicon systems is currently higher than the corresponding costs for crystalline silicon systems, since the area-related balance-of-system (BOS) costs are higher due to the lower efficiency. There is thus no cost advantage for amorphous silicon at present, and moreover the durability and long-term performance of this technology is not yet well established.

The prospects for amorphous silicon are nevertheless promising, since it is probable that the efficiency can be steadily improved to at least 10 percent. Also large-scale plants with continuous automated production are beginning to be introduced, which will achieve economies of scale. Other thin film cells are also being developed, such as Copper-Indium-Diselenide and Cadmium Telluride, which potentially offer considerably higher efficiencies, around 15 to 18 percent.

Some authorities maintain that crystalline silicon technology could remain competitive with thin film processes, after allowing for the higher efficiency and durability offered by these cells. For example, Spire Corporation (USA), a specialist supplier of crystalline silicon manufacturing equipment, maintains that a module selling price of $2.50 would be feasible with a 30MW/year plant, using advanced mono-crystalline cell technology. Even lower prices were projected by Research Triangle Institute in a study for the US-DOE in 1985, which concluded that a 25MW/year plant would be able to produce mono-crystalline modules for $1.90/Wp. This low crystalline

silicon price was independently confirmed at the ISES Solar
World Congress held in Hamburg 1987, when a leading French
industry expert said that a direct cost of $1.20/Wp at
module level would be possible with a large plant without
the need for any major technical breakthrough. A selling
price of less than $2/Wp would be sufficient to cover
indirect costs and profit.

It is of course the prices for complete systems, and not
simply the prices for modules, that are of key importance
for economic appraisal and associated market forecasts. An
important element is the BOS cost, which is in large measure
dependent on the module efficiency, since this parameter
governs array area and related costs. BOS costs for grid-
connected systems are currently about $600/m$^2$ for area-
related costs. For crystalline silicon modules costing
$5/Wp with efficiency 10 percent, the total BOS costs are
$9/Wp, giving total system costs of $14/Wp. For amorphous
silicon modules costing $4/Wp with efficiency 5 percent, the
total BOS costs are $15/Wp, giving total system costs of
$19/Wp, considerably higher than for crystalline silicon.

As more experience is gained in designing and constructing
systems, and as the volume of production increases, both the
module and the BOS costs are expected to fall significantly.
Burgess et al  (Ref. 1.6) showed that the installed cost of
grid-connected systems could come down to $2.35/Wp given a
module price of $1/Wp, area-related BOS costs of $70/m$^2$ and
power-related BOS costs of $0.63/Wp.

A projection to the year 2000 of the prices of modules (ex
works) and systems (installed) is shown in Fig. 1.9, based
on studies reported by Starr and Hacker (Ref. 1.7). The
projection distinguishes between the price trends
anticipated for crystalline silicon technology and those for
thin film technologies, with a wide band given for system
prices, depending on the application and whether stand-alone
or grid-connected. At this stage, it is hard to be sure
which technology will turn out to be the cheapest for power
applications. In the short term, it is likely that
amorphous silicon will continue to be the preferred
technology for consumer products, whereas crystalline
silicon will continue to be the preferred technology for
power applications.

1.3.3 Market Prospects

Market prospects are largely dependent on prices of
photovoltaics in relation to alternative energy sources but
other factors are important, such as official incentives,
availability of finance, environmental issues and the
general perception of the technology held by potential
customers. Although it is not possible to predict with
precision what the future market will be, Table 1.5

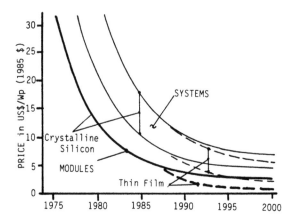

Figure 1.9      Price projection for photovoltaic modules and
                systems to 2000 (based on Ref. 1.7)

Table 1.5       Projections of photovoltaic prices and sales
                1987-2000 (1987 US$)

| Year | LOW PRICE SCENARIO | | | HIGH PRICE SCENARIO | | |
|------|---------|---------|--------|---------|---------|--------|
|      | Modules | Systems | Sales  | Modules | Systems | Sales  |
|      | $/Wp    | $/Wp    | MWp/yr | $/Wp    | $/Wp    | MWp/yr |
| 1987 | 5.5     | 10-20   | 32     | 5.5     | 10-20   | 32     |
| 1990 | 3.0     | 6-10    | 80     | 4.0     | 7-11    | 50     |
| 1995 | 2.0     | 3-7     | 300    | 3.0     | 6-10    | 100    |
| 2000 | 1.5     | 2.5-5   | 500    | 3.0     | 5-9     | 250    |

indicates what the future sales of photovoltaic systems might be, for two scenarios.

The low price scenario assumes that large area thin film cells (such as amorphous silicon) with efficiency and durability suitable for power applications become generally available within the next two or three years. The high price scenario is based on the assumption that the technical targets for thin film cells remain elusive, leaving crystalline silicon as the dominant technology for power applications. There will continue to be a growing market for consumer products powered by small area amorphous silicon cells, even if prices remain at today's levels.

For the low price scenario, with thin film module prices falling to around $1.5/Wp, total annual sales are projected to grow rapidly, from the current level of about 32MWp to as high as 555MWp by 2000, with continued expansion thereafter. Most of the output would be in and for developing countries for rural electrification and irrigation pumping, using stand-alone systems, but there would also be many applications in industrialised countries without adequate solar resources for consumer systems, professional systems and remote houses and villages. Grid-connected applications could begin to become a significant market in some countries by the late 1990's. For the high price scenario, with crystalline silicon module prices falling to about $3/Wp and thin film cells not viable for power applications, the total market would grow much more slowly, possibly reaching about 250Mwp/year by 2000. Most of the sales would be for consumer systems and professional systems, with relatively little going to rural electrification, because of the high capital costs involved. In industrialised countries with adequate solar resources, there would be good markets for isolated houses and for consumer systems, particularly for the tourist and leisure markets. Systems installed by national governments and public utilities would be mainly for applications with high social value. Although this market would be relatively limited, probably only a few megawatt per annum, the benefits to the isolated communities affected would be high.

### 1.3.4 Market Development

There is a growing commercial market for photovoltaic systems for consumer goods and professional systems in both industrialised and developing countries. Continued growth in these areas is expected, but the widespread introduction of photovoltaic systems for applications where the market prospects are very large will not begin until costs are considerably reduced below present levels. There are however good prospects for continued cost reductions and eventually photovoltaics may become a major energy resource for many countries in the "solar belt", and to a lesser extent those further north and south.  Meanwhile, as costs come down, the range of applications for which photovoltaics provide the most cost-effective solution will steadily expand.

In Europe and other industrialised regions, the markets will be largely for consumer products and professional systems. Since most of the population in these countries is already served by the grid, opportunities for electrification of islands, isolated houses and villages are limited, but nevertheless such applications are important in view of the social benefits that would follow.  It would be desirable for new buildings in these areas to be designed expressly for powering by photovoltaics, with the array integrated into the roof structure and with low-energy appliances.

In developing countries, there will be a growing market for consumer products for the wealthier sections of society, but there are much larger markets in the short and medium terms for professional systems, particularly for telecommunications, village water supplies, police posts and health centres.  If stand-alone system costs can be brought down to about a third of current levels (ie, to around $6/Wp), which may well be possible within five to ten years, rural electrification using photovoltaics will become a viable option in many situations, with markets reckoned in many hundreds of megawatts per annum.

## 1.4 Obstacles to Progress

Until such time as photovoltaic systems become cheap enough to be economically viable for grid-connected applications, the markets in industrialised countries will remain relatively limited.  Apart from the cost factor, there are no major technical or institutional barriers to be overcome, although there is still a need for information on photovoltaics to be disseminated to potential customers, many of whom remain unaware of the opportunities offered by this relatively new and formerly exotic technology.

In developing countries, there are still many obstacles to be overcome.  There is a real danger that the widespread introduction of photovoltaics for applications where the benefits are high and the market prospects are very large indeed, principally systems for water pumping and rural electrification, will be held up for three main reasons:

i)   The costs remain too high;

ii)  The majority of the hardware has to be imported;

iii) There is inadequate institutional support.

Each of these factors is discussed below.

## 1.4.1 High Costs

Photovoltaic systems are currently too expensive for many applications to be viable. Although a photovoltaic system may have attractive operational advantages compared with the alternatives, its high cost rules it out as a commercial proposition. Furthermore, many potential customers are holding back since costs are predicted to continue falling over the next few years.

The problem here is that the predicted cost reduction partly depends on technical development and partly on increased scale of production. If buyers hold back until costs come down, manufacturing volume cannot be increased and research and development is starved of funds. This is the reason why only manufacturers with substantial financial resources are able to survive, and why public funds continue to be needed to support research, development and demonstration activities.

## 1.4.2 Local Manufacture

Clearly, if photovoltaic systems are to become a significant energy resource for a country, its government will wish to build up its own manufacturing capability and not be dependent on imports. This is already happening in some countries, notably Brazil, India, Mexico and China. For other countries, manufacture of photovoltaic cells would not be practicable for many years, particularly in view of the uncertainties regarding the best thin film technology, but nevertheless local assembly of modules using imported cells could well be feasible, with local components being used wherever possible for the balance of systems.

In most cases it would be an advantage for the local industry to have the benefit of the design experience and manufacturing know-how of an established photovoltaic company, through some joint venture or licensing agreement. Independent advice from specialist consultants would also be advisable, to ensure all alternatives are properly considered.

## 1.4.3 Institutional Development

In the industrialised countries, established institutions already exist for funding photovoltaic research, development and demonstration and for informing the public of the potential costs and benefits of the new technology. In the developing world, a major need in many countries is for an effective institutional base that can monitor, plan, coordinate and regulate developments in a technology as new and promising as photovoltaics.

Public and private interests need to be brought together to ensure that each of the following aspects is properly covered:

a)    Technical Development

-    identification of projects that are properly justified on technical, economic and social grounds
-    applied research to develop appropriate systems for local needs
-    commercial manufacture and assembly of components and systems
-    integration of photovoltaics into related fields (eg, building construction and electronics)
-    field installations for demonstration purposes, with monitoring and evaluation

b)    Regulatory Aspects

-    development and implementation of appropriate codes and standards
-    regulations for consumer protection and warranties
-    independent testing and certification of components and systems
-    planning of integrated projects with no disturbing side effects

c)    Incentives and Finance

-    tax credits, subsidies and grants
-    coordination of aid-funded projects
-    public and private finance for manufacturers and customers

d)    Training and Information

-    training of professionals in all aspects of photovoltaics
-    public information and advisory services
-    liaison with other energy supply agencies

An appropriate institutional base to deal with the above tasks is vital if a country is to protect its interests and derive maximum benefit from the emerging photovoltaic technology.    The international aid agencies and the established photovoltaics industry can do much to encourage and assist this process.

1.5 Conclusion

Space technology has indeed come down to earth.    No longer an exotic, highly expensive technique for very specialised applications, photovoltaic systems are now a practical proposition for many off-grid applications.    Research and development efforts are continuing on a broad front worldwide to reduce costs still further. As this happens, the number of applicants which are economically viable will steadily increase. Photovoltaic systems serving remote communities, although at present quite expensive, offer a number of advantages, since they require no fuel and very little maintenance. They also give rise to no harmful pollution.

In time, given further technical progress, photovoltaics could well prove to be a significant energy resource, generating electricity for the grid alongside conventional plant, as well as providing stand-alone power for remote houses and villages. Meanwhile, there are many consumer and professional applications today which are finding growing markets in both industrialised and developing countries.

## 1.6 References

1.1 Kenna J P and Gillett W B 1985 <u>Solar Water Pumping - a Handbook</u> (London: Intermediate Technology Publications)

1.2 Burgess P J and Prynn P J 1985 <u>Solar Pumping in the Future - a Socio-economic Assessment</u> (Cardiff: CSP Economic Publications)

1.3 Starr M R 1985 <u>Photovoltaic Prospects for Rural Electrification</u> Proc. INTERSOL 85 Solar Energy Congress, Montreal, p1698, [Pergamon Press; 1985]

1.4 Starr M R 1987 <u>Photovoltaics for Rural Electrification</u> Proc. First International Conference on Power Sources and Supplies, London, April 1987 (Electrical Research Association)

1.5 Maycock P D 1987 <u>Photovoltaic Market Forecasts</u> Photovoltaic News, Vol. 6. No. 11., November 1987

1.6 Burgess E L. Biringer K L and Schueler D G 1981 <u>Update of Photovoltaic System Cost Experience for Intermediate Load Applications</u> Proc. 15th IEEE PV Specialists Conference, Orlando, p1453, [IEEE; New York, 1981]

1.7 Starr M R and Hacker R J 1985 <u>Updating the Cost Projection for PV Modules and Systems</u> Proc. Sixth EC Photovoltaic Solar Energy Conference, London, p470 [Reidel; 1985]

# Photovoltaics for Development: Identifying Real Opportunity

Phil O'Keefe

Dept. of Environment, Newcastle Polytechnic,
Ellison Place, Newcastle Upon Tyne, NE1 8ST

## 2.1 Introduction

This chapter draws upon recent experience in Third World
energy planning. It argues that successful diffusion of
photovoltaic systems depends upon a clear identification of
the specific niche of PV systems based upon enduse analysis.
Such an analysis clearly indicates that, whilst their
contribution to energy budgets will be small, PV systems
present an ideal development opportunity.

Over the past decade, several attempts have been made to
provide a "Bird's Eye View" on energy possibilities in the
Third World. The view from the clouds, however, is very
different from the ground view. A ground view a "Worm's Eye
View", is necessary to counter-balance the "Bird's Eye
View". In this chapter, we will outline such a view,
teasing out the implications for the development of
photovoltaics and its place in the national energy policy
options for developing countries.

The chapters preceding and following this one concentrate
exclusively on describing appropriate applications of PV
systems. This chapter attempts to outline a conceptual
framework within which the role of photovoltaics can be
discussed. It is as important to understand what the
technology cannot do (or, at least, cannot yet do) as it is
to promote those applications for which it is particularly
well suited. The examples given of socially inappropriate
use of solar energy for cooking or the misunderstanding of
the role of woody biomass serve as warnings to those
promoting photovoltaics or other new energy technologies.
The social context is a crucial component in the
introduction of any new technology and failure to consider
this from the outset will lead to a condemnation of the
technology and alienation of those who could have benefited
from its use.

## 2.2 Third World Energy

Third World energy is dominated by two problems. In the
commercial sector, oil is the predominant fuel; in the
traditional sector, biomass, especially in the form of wood,
dominates consumption. Oil is critical for industry, but
more importantly for transport. As little capital is
available for the development of public transport systems,

eg. railways, oil consumption will continue to expand over the next twenty five years, although the impact of the Third World demand on global energy markets will not be substantial. Biomass demand will expand since biomass is the predominant fuel for household energy in both rural and urban areas. The nature of this expansion will be driven by demographic change, namely the overall population growth rate, the rate of rural to urban migration and the rate of household formation. Since much of the demand for traditional sector fuels lies beyond commercial markets, it is impossible to provide conventional analysis of fuel substitution using income elasticity; demographic change, therefore, serves as a crude measure of future demand.

Biomass dominates national energy balances. In Africa, south of the Sahara, biomass accounts for between 95% and 50% of total energy use. In Tanzania, and throughout the Sahel, the energy consumption is towards the upper end of this range. Zimbabwe, the most developed economy in independent Africa, represents the lower end of the range. In all areas, however, the enduse which dominates is cooking, accounting for some 80% of household demand. There is growing concern, notably with the accelerated rate of urbanisation, that wood will be increasingly drawn from stock not yield. Such a process would increase problems of environmental degradation.

In reviewing the Third World energy situation and suggesting a future tense, the twin pillars of energy consumption remain oil and biomass. Coal will find increasing use, particularly in industry, but the heavily promoted New and Renewable Sources of Energy (NARSE) will have little impact on national energy budgets. Conservation opportunities, too, are frequently limited, not least because of the lack of capital for re-equipment.

An emphasis on efficient engineering and demand management has produced new forms of energy appraisal. Central to this appraisal is the identification of enduse demand, broken down by fuel, technology and intensity of use as well as indicating the social class of user, for specific purposes. It is the critical tool for disaggregated energy analysis and a critical tool that can be used to identify the niche for photovoltaic deployment.

## 2.3 Enduse Analysis

Enduse analysis is useful because it can begin to identify the constraints on the diffusion of any new fuel-technology combinations. A useful illustrative example of this is the current experience with the diffusion of improved stoves. Traditional stoves, including open fires, are notoriously inefficient. As cooking dominates enduse activity in the Third World, substantial energy savings would seem to be

available with improved design.   The problem, however, is
that traditional stoves provide a range of energy services
that preclude rapid diffusion of improved stoves.

In modern society, fuel-technology combinations are specific
to enduse.   In the Third World, there is a simultaneity of
enduse.   A traditional stove encompasses a power range for
simmering and boiling in addition to serving as a source of
space heating, food drying, insect control and a social
focus.   Most importantly, it is the major source of light.
Improved   stoves,   largely   improved   by   confining   the
combustion space, limit these other simultaneous enduses,
notably light production.   It is not surprising therefore,
that there has been limited success with new stoves not only
because they are more expensive but because they do not
provide alternatives for the simultaneous enduses that
disappear with increased stove efficiency.

A similar example can be drawn from fuelwood projects.
Because fuelwood dominates the national energy balances of
many developing countries and because, simultaneously, there
was increasing concern about deforestation, there was a
major push to develop fuelwood projects.   Despite spending
over US$ 1,000,000,000 over the last fifteen years on such
initiatives, the global returns have been marginal.   Why
should this be so?

Firstly, while it is true that fuelwood dominates energy
balances,   the   balances   themselves   are   "incorrect".
Electricity, gas, coal and oil are primary or derived fuels,
i.e.   their   dominant   characteristic   is   as   an   energy
commodity.   Fuelwood, however, is a residue from other uses
of wood: fuelwood is what is left over when a tree ceases to
be useful for food, fruit or fodder.   Fuelwood is, in short,
rubbish.   And, even if this rubbish dominates the energy
balance, is it possible to devise a system for commercially
growing   it   when   its   traditional   forms   of   production,
collection and utilisation are largely beyond the market?

Secondly, although fuelwood dominated the enduses of all
wood, there is no direct linkage between fuelwood usage and
deforestation except where urban demand leads to the
commercialisation of fuelwood trading and consequently,
especially with charcoal production, to the cutting of whole
trees.   In most rural areas, people harvest wood yield not
stock   through   a   range   of   pollarding   techniques.
Deforestation, when it occurs, is largely the result of land
clearance for agriculture, which ironically produces a
fuelwood surplus even if the surplus situation is short
lived.

Thirdly, and perhaps most importantly, fuelwood supplies
come not from forests but from trees outside the forest.
Forests are largely monocultural plantations in marginal

areas. As such, they are remote from population centres, distant from the sites of consumption. People, however, obtain supplies from close to the homestead, from their own fields, from hedgerows and roadsides and from common land. And yet, with one notable exception (Ref. 2.1), there is no knowledge of the quantity and spatial distribution of this resource. Moreover, the measurement of forest productivity, largely concentrates on stem volume not total woody biomass production, while people consume fuelwood not as tree trunks but as branches and twigs.

Finally, the foresters themselves, trained as rangers (environmental policemen) or an engineers, have limited sympathy with the complex intercropping practice of peasant farmers. Conversely, agriculturalists have little understanding of the characteristics of trees and frequently remove all woody biomass in opening new land for arable production.

This misunderstanding of the nature of the fuelwood problem is compounded by professional shortsightedness. The solution rapidly becomes the problem. Foresters move their monocultural technologies from the forest to the village to begin "community forestry" or "social forestry" without understanding community perception or social needs. The problem is defined as a "fuelwood" crisis so "fuelwood" trees are emphasised. The failure of these initiatives is blamed upon local custom or gender relationships when the failure is essentially rooted in a misunderstanding of the multidimensional, multipurpose, multifunctional use of wood in Third World rural communities.

Ironically, in both the conservation (stoves) and supply enhancement (fuelwood) strategies, the technologies are largely proven. It is not a technical problem but a problem of matching social needs to technology. Crudely, it is a problem of understanding the purpose of energy within the social structure. And what is problematic for traditional energy patterns is equally true for the New and Renewable Sources of Energy (NARSE).

NARSE is increasingly presented in an ideological fashion. For example, many texts on the subject advocate a demand-driven approach to the energy problem but then reverse logic and present NARSE technologies as supply solutions. Such logic reversal is spectacularly obvious with some solar technologies, notably solar cooking.

The failure of solar cookers rests not on technical considerations but upon a misunderstanding of the place of cooking within the traditional division of labour. Women, who are responsible for cooking, have a large labour contribution both to agriculture and household tasks. Consequently, they are not free to use such cookers until

early evening, by which time solar cookers are relatively
useless.  Enduse analysis incorporating such social analysis
would have provided an early indication of probable failure.

## 2.4 Towards a Structural Model for Photovoltaic Diffusion

There is little doubt that, of all the New and Renewable
Energy Technologies, photovoltaics offer the most promise.
The difficulty, however, is to gauge the niche of such
technologies accurately.  The old adage that: "To Know is to
Adapt" is clearly inapplicable not least because it does not
provide an explanation of energy change, over-emphasises the
importance of information and over-simplifies the barriers
and resistance to change in the real world.   It is
impossible to provide a model for energy diffusion that is
abstract and devoid of social context.  Meaningful analysis
must rest upon an assessment of individual technologies
within a specific social context given identified project
goals.  Three broad areas of rural energy development can be
identified, namely;

* Energy for Household Consumption
* Energy for Agricultural Consumption
* Energy for General Development

Within this context, and with a knowledge of the three broad
areas of energy demand, it is possible to ask a series of
questions to evaluate the social opportunity offered by
NARSE.   These questions follow the logic of a project
assessment cycle which include:

* Problem Identification
* Preliminary Specification of Project
* Detailed Data Collection
* Project Analysis
* Evaluation

Figure 2.1 contains a schematic expression of this project
assessment cycle.

Once a problem, a broad need, has been identified - although
the definition of the problem is an iterative process
throughout the project assessment cycle - these types of
information are needed for a preliminary specification of
the project, namely:

    1. Technical parameters
    2. Project site and situation (locale)
    3. Project goals

This information is collected with reference to single or
multiple users so that, from the beginning of the project
cycle, the endusers are clearly identified and the scale of
the project is specified.

This preliminary specification is the basis for detailed data collection on the social structure, the system into which NARSE technology will fit. There are five broad areas of investigation:

1. Resource Ownership or Resource Access
2. Institutional Frameworks (both public and private)
3. Existing Infrastructure
4. Labour Availability
5. Socio-Cultural Values

These broad areas of investigation will allow the construction of indices that enable the preliminary specification to be analysed. These indices address issues of:

AFFORDABILITY What will it cost individuals or society, bearing in mind that, in the Third World, there is a simultaneous cash flow and capital shortage, which is resolved by foregoing capital investment? Classical arguments for investment in NARSE (higher capital but negligible running costs) are attractive to individuals or government only if suitable financial packages can be arranged, since capital investment expands the debt burden.

ACCESS What level of access will there be to the technology (i.e. transfer to the user countries), not least to the total engineering system? This is frequently more problematical to access than the energy supply itself.

LABOUR BUDGET
CONSTRAINTS Given that the single most important constraint on peasant society is lack of labour, what does the total engineering system mean for the reorganisation of labour? Above all, have the planners assumed that labour is free of charge and freely available?

CULTURAL
ACCEPTABILITY Does the technology fit existing preferences or is its success predicted on a major shift in cultural norms and practices?

Such questions are essential as they indicate whether or not a broader social impact model, that would provide backing for arguments seeking financial support from outside the community, is required. Future evaluation will indicate whether indices were accurate or whether they were so aggregated that they masked the pattern of probable adoption. The model, outlined in Figure 2.1, provides a useful checklist against which the project assessment cycle can be measured.

In passing, however, note should be taken of the mechanisms for diffusion of rural development initiatives. Three broad mechanisms are available, namely the market, the agricultural extension system and the health care delivery system. The market system (known as the Coca Cola Model because that dark, sweet carbonated water is everywhere) works when goods are commodified: it does not work well when goods (e.g. fuelwood) are free or when technologies are not proven and require subsidies for demonstration processes. In other words, using the market system to deploy NARSE for Third World rural development is likely to be successful only where the potential users are already part of the monetary economy. The agricultural extension system is useful for sharing knowledge and low cost inputs, but is poorly suited to high cost equipment transfer. The health care delivery system is excellent for preventive action at low cost - somewhat parallel to the energy conservation initiatives focused on individual households.

Reviewing mechanisms for diffusion would suggest that there are limited opportunities for NARSE amongst the rural poor, since only the market mechanism can address the capital costs associated with NARSE. That simple stark conclusion perhaps requires rephrasing - the diffusion of NARSE amongst the rural poor can be accomplished only if there is a movement to establish collective action and purchase as, for instance, in Mali, with collective access to capital. However, in the urban industrial sector, where energy supplies and technologies are already commoditised and already command their own markets, there are real possibilities for NARSE to be used as a fuelswitch from conventional commoditised energy forms.

## 2.5 Rapid Appraisal for Energy Projects

Many energy planners will argue that the detailed analysis laid out in the preceding section is too time consuming, too complicated to undertake. This argument must be dismissed not simply because its starting point is dismissive of local needs, but also because it frequently results in failed projects, in dismissed technologies where dismissal results from poor social analysis, not poor technology. If, however, planners seek a simplified series of questions, the list below can be used for rapid appraisal where, if the questions are answered affirmatively, investment should be withheld.

Figure 2.1 General model for the analysis of rural energy technology diffusion patterns

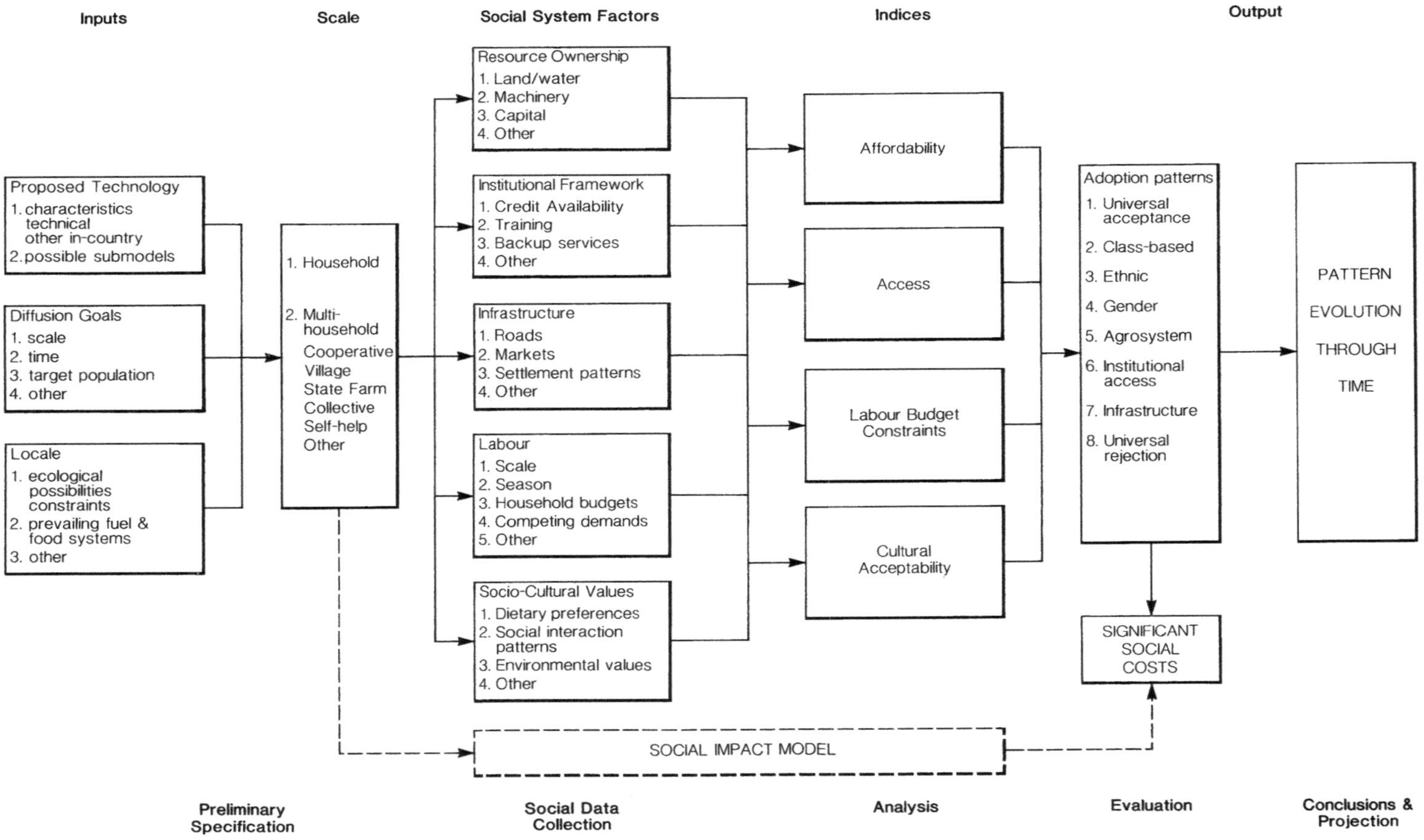

| Inputs | Scale | Social System Factors | Indices | Output |
|--------|-------|----------------------|---------|--------|

**Proposed Technology**
1. characteristics
   technical
   other in-country
2. possible submodels

**Diffusion Goals**
1. scale
2. time
3. target population
4. other

**Locale**
1. ecological
   possibilities
   constraints
2. prevailing fuel &
   food systems
3. other

1. Household

2. Multi-
   household
   Cooperative
   Village
   State Farm
   Collective
   Self-help
   Other

**Resource Ownership**
1. Land/water
2. Machinery
3. Capital
4. Other

**Institutional Framework**
1. Credit Availability
2. Training
3. Backup services
4. Other

**Infrastructure**
1. Roads
2. Markets
3. Settlement patterns
4. Other

**Labour**
1. Scale
2. Season
3. Household budgets
4. Competing demands
5. Other

**Socio-Cultural Values**
1. Dietary preferences
2. Social interaction
   patterns
3. Environmental values
4. Other

Affordability

Access

Labour Budget Constraints

Cultural Acceptability

**Adoption patterns**
1. Universal acceptance
2. Class-based
3. Ethnic
4. Gender
5. Agrosystem
6. Institutional access
7. Infrastructure
8. Universal rejection

PATTERN EVOLUTION THROUGH TIME

SIGNIFICANT SOCIAL COSTS

SOCIAL IMPACT MODEL

| Preliminary Specification | Social Data Collection | Analysis | Evaluation | Conclusions & Projection |
|---------------------------|------------------------|----------|------------|--------------------------|

1.   Technologies implying a commodification of basic needs
     or  a  privatisation  of  common  resources  often
     implemented to the detriment of poorer households.

2.   Technologies that disregard the fine balance between
     fuel provisioning and other basic needs.

3.   Technologies that replace complementaries in existing
     agro-fuel  systems  with  elements  that  compete  for
     available resources.

4.   Technologies  increasing  local  dependency  on  outside
     institutions.

5.   Technologies whose maintenance requirements cannot be
     guaranteed  either  by  local  expertise,  the  project
     proponents or national agencies.

6.   Technologies  whose  benefits  and  costs  increase  the
     (absolute  or  relative)  vulnerability  of  the  poorest
     groups.

7.   Technologies  focused  on  environmental  conservation
     practices at the expense of sustained or increased crop
     yields.

8.   Technologies which further burden already overworked
     household members such as women.

9.   Technologies requiring major changes in diet.

10.  Technologies disruptive of household routine and social
     interaction patterns.

## 2.6 Social Evaluation of PV Systems

The broad conclusions from the implementation of such an
analytical  model  with  reference  to  photovoltaic  systems
would seem to indicate the following conclusions.  Firstly,
there  is  little  scope  for  the  PV  system  purchase  at
individual  household  level,  although  collective  action  at
village  level  can  bring  access  to  sufficient  capital  to
install PV systems for drinking water and lighting.

The strength of PV systems lies in their contribution to
agriculture and general development.  Sufficient experience
has  now  been  gained  from  the  operation  of  PV  systems  to
suggest that the systems can serve important enduses (Figure
2.2).  These would include, ranked by achievement:

1.   Installation  of  telecommunications  to  accelerate
     national integration and development.

2.   Provision of clear water for drinking and medical care.

Figure 2.2     Market readiness of photovoltaics

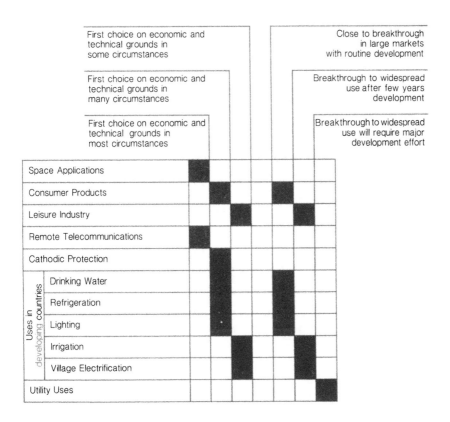

3.  Expansion of pump irrigation without which it will be extremely difficult to meet the expanded food requirements of the Third World.

4.  Provision of refrigeration facilities, in medical institutions, for the promotion of primary health care.

5.  Lighting.

Third World telecommunications, the responsibility of government institutions, frequently utilise PV systems. The current benefits include modularity, high reliability, low maintenance and no fuel distribution costs. The high reliability of PV systems produces increased revenue to the telephone network. The cost effectiveness, in this case, is wider than a simple price/Watt comparison with other sources of power.

High quality potable water is a major determinant of the quality of life not least because it allows control of endemic disease. In many senses, the single greatest technical challenge to alleviate Third World poverty is to raise water some 20 metres. This is particularly so in remote areas. Again, the major contractor is a government body. Although there is a reluctance to charge consumers for potable water, there is clearly a market niche for safe drinks. The global expansion of Coca Cola could not otherwise be explained! The direct benefit to government is the enhancement of its people's quality and quantity of life.

Irrigation and watering of animals is another successful application. Secondary benefits include expanded indigenous food production and a consequent drop in food imports. Direct primary benefits usually accrue to individual farmers. These farmers are usually already involved in a cash economy but initial capital costs of PV pump systems require the establishment of additional financial packages to support equipment purchases.

The maintenance of the cold chain to preserve vaccines requires refrigeration. This is again the responsibility of government agencies. The PV purchaser is technically sophisticated and requires efficiency and reliability - the critical economic parameter for cost effective vaccination. For isolated medical centres, PV refrigeration is an attractive proposition.

There is a wide range of customers for PV lighting systems. As such, customers have different standards and requirements. Government agencies are critical purchasers because they supply lighting services to remote areas.

It is useful to separate the issue of lighting, a necessary service to expand the length of the working day, from the broader issue of rural electrification. In the latter case, PV systems for varying peak loads can have high costs, although for baseload provision it is attractive in remote isolated areas.

Social evaluation suggests that PV systems will have little impact on household energy consumption, the largest sector of energy consumption in the Third World, since they will have little impact on cooking, the single most important enduse. In short, PV systems will have little impact on national energy budgets in general. Such conclusions, however, disguise in national statistics the critical role that PV systems can play in development, raising productivity in agriculture and the quality of life in general.

## 2.7 Conclusion

Proponents of new technology have always taken a strong supply-side approach to problems. In energy problems, where costs frequently bear little relationship to real price, this tendency is particularly marked. Proponents of PV systems need to evaluate carefully the niche of their system on the basis of demand, not supply. Such an evaluation would suggest that these systems have significant development opportunity although their contribution to energy supply is minimal. It is, however, development that will contribute to the easing of the household biomass energy crisis.

The 1985 shipment of photovoltaic modules was 24.4 MW worldwide, excluding the Eastern block countries for which no figures are available. This represents a market for photovoltaic modules of over £125 million. In addition, the market for associated batteries and control systems will match or exceed this figure. This market, however, is not all embracing. The common perception is that the market will reach a level of about 500 MW per annum by the mid 1990's and over 1 GW by the turn of the century.

PV systems, notes the worm, are too important to be dismissed by over-selling. Their role, in a strategic assemblage of energy strategies is critical, not least because there is not one single energy problem in the Third World. There are many energy problems requiring many solutions. PV systems have a critical role in the provision of specific energy services which promote social and economic development, although their contribution to the national energy budget may be small.

2.1 Environment and Development in Africa, Vols.1-6, Ed. J.T.C. Simoes, [Beijer Institute and Scandinavian Institute of African Studies, Sweden; 1984].

# Photovoltaics for Developing Countries

Bernard McNelis

IT Power Ltd.,
The Warren, Bramshill Road,
Eversley, Hants.

## 3.1 Introduction

Energy is needed for practically all the activities that are basic to human survival, such as cooking, water pumping and food production. After basic needs are satisfied, further energy is required to improve the quality of life, through lighting, transport, telephone communications and consumer tools such as refrigerators, radios and televisions. As a country develops, still further inputs of energy are required for industries and for commercial and public buildings. In urban areas, the necessary energy supplies may be readily provided through oil products, coal and networks for electricity and natural gas. In rural areas, traditional sources of energy, principally firewood, agricultural residues and cattle dung, continue to be of major importance, supplemented by commercial sources such as electricity and oil products in areas where the physical infrastructure makes this possible.

The majority of the population of all developing countries live in the rural areas. The combined effect of population growth and supply problems of commercial fuels is putting ever-increasing pressure on the traditional fuel supplies. Deforestation resulting from over-cutting of trees, sometimes aggravated by long-term climatic changes, is becoming a major problem in many countries. The use of agricultural residues and cattle dung as fuel reduces the amount of nutrients returned to the soil.

Photovoltaic systems are widely recognised as an attractive means to address some of the rural energy problems, since they offer the following advantages:

* Being built up from solar cell modules, they are able to provide relatively small amounts of electrical power at or close to the point of demand.

* No fuel requirements.

* Relatively simple operation and maintenance requirements, within the capability of unskilled users.

* No harmful pollution at the point of use.

* Long life with little degradation in performance.

Some of the principal applications are discussed below.

## 3.2 Water Pumping

### 3.2.1 Pumping Techniques

Hand and wind powered pumps have a long history of lifting water in rural areas for water supply and irrigation. Improved designs of hand pumps that are more efficient and durable and easier to maintain are now being widely introduced. There has also been renewed interest in wind pumps in recent years, with the emphasis on lower cost designs suitable for local manufacture. Petrol and diesel pumps are also being used in some areas, particularly for low lift irrigation.

In rural areas that have been electrified by grid extension, electric pumps are usually a reliable and relatively low cost option. However, for most rural areas, it will be many years before this alternative is available.

Substantial efforts have been made in recent years to develop reliable and cost effective solar powered pumping systems. A number of prototype solar thermal systems have been developed, but none so far offer sufficient reliability, ease of operation and maintenance and cost effectiveness. Photovoltaic systems, on the other hand, offer a number of attractive features and, after several years of development, are now readily available in various standard configurations, as shown in Figure 3.1.

A solar photovoltaic (PV) water pumping system consists of the following main components: the PV array, with support structure, wiring and electrical controls; the electric motor; the pump; and the delivery system, including pipework and storage. These components have to be designed to operate together to maximise the overall efficiency of the system (or, rather, to optimise the cost effectiveness of the system). An electrical controller is sometimes incorporated to improve the electrical performance of the system. Energy storage in the form of batteries is rarely used, as it is generally cheaper and simpler to store the water to cover periods of low solar input or high demand.

The advantages and disadvantages of the various pumping techniques are compared in Table 3.1. The main problem with PV pumps has been their high initial cost, but with cheaper PV modules coming onto the market and with improved system designs incorporating volume-produced pumpsets, this does not constitute such a barrier.

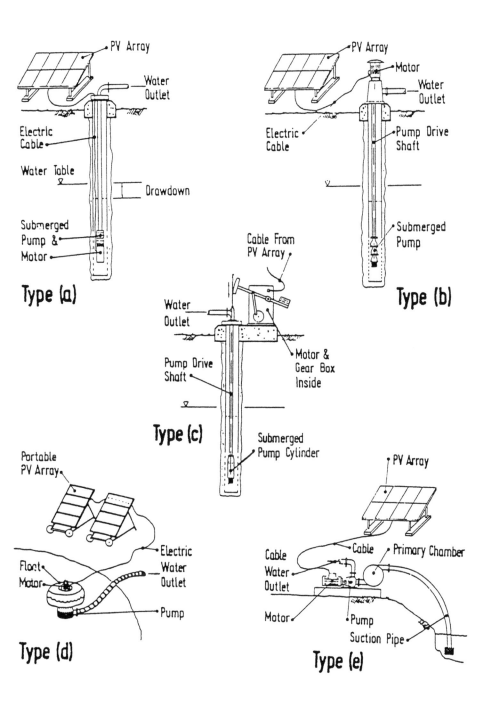

Figure 3.1    PV Water Pump Configurations

Table 3.1  Comparison of Pumping Techniques

| Pumping Technique | Main Advantages | Main Disadvantages |
|---|---|---|
| Hand pumps | Low cost. Simple technology. Easy maintenance. | Low flow. Absorbs time and energy that could be used more productively elsewhere. Often involves uneconomic use of expensive bore-holes. |
| Diesel and gasoline pumps | Low capital cost. Can be portable. Extensive experience. Easy to install. Easy to use. | Maintenance often inadequate, reducing life. Fuel often expensive and supply unreliable. Noise, dirt and fume problems. Unreliable if not maintained |
| Wind pumps | Moderate capital cost. Suitable for local manufacture. Easy to maintain. Non-polluting. Needs no fuel. Long life. Extensive experience. | Very sensitive to wind speed, with periods of low output. Needs open terrain. Not easy to install. |
| Solar PV pumps | Low maintenance. Non-polluting. Needs no fuel. Easy to install. Reliable. Long life. System is modular. | High initial cost. Low output in cloudy weather. |

### 3.2.2 Water Supply

An example of a solar photovoltaic borehole pump used for village water supply is shown in Figure 3.2. Water supply requirements do not vary much month by month. It is important to provide sufficient storage to cover periods of cloudy weather, when the output from the PV pump will be low. A covered tank at or near ground level, connected to a number of automatic shut-off supply taps on a concrete or stone pad would be a typical arrangement.

Whenever possible, it is safer to take water intended for human consumption from enclosed boreholes or protected wells. If a surface water source, such as a lake or a stream, has to be used, it is usually possible to construct some form of filter when building the pump sump. Complete solar powered water treatment plants are now available, as discussed later in this chapter.

Water intended for livestock is usually pumped from a borehole and stored in a raised tank so that the cattle drinking troughs may be gravity fed through ball valves. The PV array needs to be well protected to prevent damage by livestock.

### 3.2.3 Irrigation

PV pumps are well suited to irrigation applications. They produce the most water when the solar radiation is greatest and hence when the crop water demand is highest. Because PV pumps deliver water over a period of about 10 hours each day, it is important to plan carefully the distribution of the water to avoid losses by evaporation and infiltration. The irrigation technique will need to be adapted to take best advantage of the available water. For example, instead of one large pump, it may be better to deploy several small pumps at different places in a large field, or in several separate fields. Alternatively, it may be feasible to store water for discharge at a higher flow rate over a shorter time.

The pumping of relatively large volumes of water for flood irrigation for rice is unlikely to be cost effective, whereas a fruit farmer may well find that the relatively small volumes required for a trickle system could be supplied very economically. The underlying principle is that the cost of water used must be less than the value of the extra crop gained through the irrigation.

A typical PV irrigation pumping system is shown in Figure 3.3.

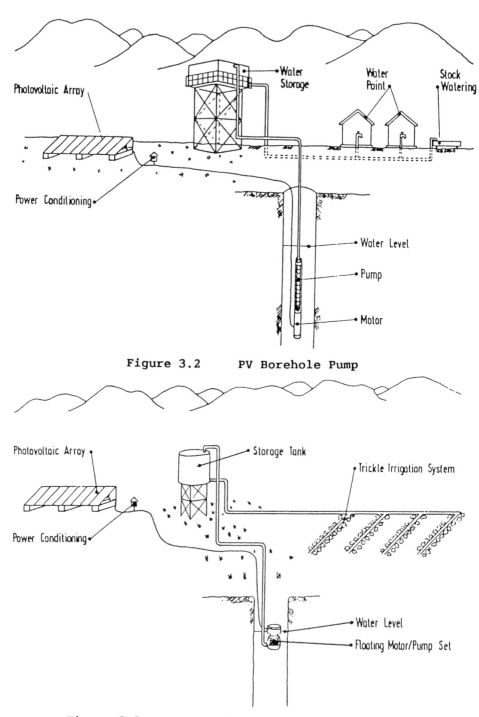

**Figure 3.2      PV Borehole Pump**

**Figure 3.3      PV Irrigation Pumping System**

3.2.4 Field Experience

A substantial volume of field experience is now available relating to solar pumps. Approximately 4,000 units made up of the different configurations shown in Figure 3.1 have been supplied worldwide. Most of these are for water supplies for villages and livestock watering, but some have been installed primarily for irrigation. A comprehensive study of solar pumps, involving field and laboratory testing of component and complete systems, was completed in 1983 by consultants for the UNDP and World Bank (Ref. 3.1). A survey of solar pumping field performance was carried out by consultants for the World Bank in 1986 (Ref. 3.2). There is also over ten years of experience available with Mali Aqua Viva, who have sponsored some 50 PV pumps, mainly for village water supply (Ref. 3.3). The experience to date in many parts of the world has been reviewed in a number of reports (Refs. 3.4-3.11).

The performance of many early systems was disappointing due to a number of factors, including:

-   Use of unreliable and/or inefficient sub-system components (motors, pumps and power conditioning equipment)

-   Poor overall system design, resulting in poor matching between the components in relation to the solar input and water level

-   Use of inaccurate data regarding solar input and water resource conditions at the design stage.

There is evidence that the manufacturers have learned from past experience and that the latest types of pumping systems are considerably more reliable and efficient than earlier models.

3.2.5 Conclusions

3.2.5.1 Technical Aspects

PV pumping technology has improved significantly over recent years, with the emphasis on better matching of system components, increased reliability and reduced maintenance requirements. The type and size of system needs to be chosen carefully on the basis of a systems approach to the problem, taking into account all relevent factors, including operation and maintenance implications. (See Ref. 3.7 for design and procurement recommendations.)

PV arrays based on the well-proved crystalline silicon cell technology have generally proved to be the most reliable component of a pumping system. With the advent of lower cost thin film cell technologies, it will be necessary to monitor array performance carefully to ensure adequate provision is made for long term degradation.

The introduction of brushless DC motors by some manufacturers for surface mounted or floating pumps has eliminated the need for brush replacement. For submerged motors, water-filled AC induction motors are proving to be much more reliable than sealed DC motors. This type of borehole system is now preferred to turbine pumps with surface motors and long vertical drive shafts. The variable frequency DC-to-AC inverters required for AC systems provide a low cost means of matching the PV array output to the motor load and they appear to perform reliably in the field.

Centrifugal pumps can be well matched to PV arrays. Problems have been reported with surface mounted centrifugal pumps, due to the need to maintain prime. A self-priming tank on the suction side has proved to be more reliable than a foot valve. Centrifugal pumps should not be used for suction lifts more than 5-6 m and wherever possible a floating or fully submerged unit is to be preferred.

Positive displacement pumps have a water output that is practically independent of head and directly proportional to speed. For PV powered systems, problems have been experienced due to the cyclical nature of the load on the motor and the high frictional forces, particularly at start up. At high heads, this type of pump can be more efficient than a centrifugal pump, since the frictional forces are relatively small compared with the hydrostatic forces. Positive displacement pumps are, however, usually very rugged and reliable, provided the overall system has been well designed in the first place to suit the conditions obtaining at the site.

A common cause for pump and/or motor failure has been overloading due to sediments in the water or tight shaft bearings. Dry running due to loss of prime (surface pumps) or falling water level in the well is another common cause of failure. Increasingly, manufacturers are providing low water level and/or high temperature protection for the motor.

The use of tracking PV arrays, maximum power point trackers and batteries may offer advantages in theory but experience has shown that at remote sites where maintenance is difficult to provide, the extra complexity introduced is counter-productive.

Pump performance is heavily dependent on the assumptions made at the design stage regarding solar and water resource characteristics. Careful account has to be taken of the variations in solar input to the array, the static water level in the well and the water demand. Failure to do this has resulted in many systems being undersized so that they fail to meet the demand, or excessively oversized, with associated additional capital cost. However, these problems should diminish as experience builds up and a larger data base applicable to each country becomes available to system designers and suppliers.

### 3.2.5.2 Economic Aspects

The F.O.B. prices of PV pumping systems have been steadily falling from about $30/Wp in 1978 to as low as $10/Wp in recent years. To this has to be added the cost of shipping and installation. The unit water costs expressed per volume-head product ($/m$^4$) may be calculated on a life cycle cost analysis for different assumptions regarding demand, solar insolation and fuel costs. This has been done in Ref. 3.12, the results of which are presented in Figure 3.4. The data assumed in the analysis is given in Table 3.3. It can be seen that solar pumps are typically competitive up to 1000 m$^4$/day demand (e.g. 40 m$^3$/day pumped 25 metres). This approximates to 1400 Wp array power. Ref. 3.2 also concluded that PV pumps are competitive up to approximately 1000 Wp (and, for regions of very high diesel operating cost, up to 2.5 kWp). Wind pumps would probably be cheaper than PV pumps if the mean wind speed in the periods of maximum water demand is at least 2.5 m/s.

Water for irrigation is characterised by a large variation in demand from month to month. Hence, a solar pump sized to meet the peak demand is under utilised in other months. This adversely affects the economics/unit water costs and it is for this reason that solar pumps are more competitive for rural water supply.

Even though the unit water cost for PV pumps may be cheaper than for diesel pumps in certain circumstances, this does not necessarily imply that PV pumps would be an economic solution for irrigation applications, since the economic market price of the crop has to be considered. Nevertheless, PV pumps combined with trickle irrigation or other low water use irrigation techniques, when used for fruit and other high value crops, may be found to be economic. Finance in the form of capital grants and low cost loans will be needed to bring high capital cost/low running cost equipment within reach of the potential users.

### 3.2.5.3 Social and Institutional Aspects

PV water pumps have found wide social acceptance, particularly in villages which previously had to pump water by hand. However, PV pumping systems, being a new technology, need continuing institutional support to enable

*Applications of photovoltaics*

Table 3.2 Typical Cost Analysis of PV Pumping System in Mali.

**BASIC DATA**

| | |
|---|---|
| System size | 1400 Wp |
| Capacity | 160 m³/day at 5m head |
| Hydraulic power | 361 W |
| Use(assumed availability) | 340 days/year |
| Annual output | 54400 m³ |

**COST ANALYSIS**

| | |
|---|---|
| Period of analysis | 15 years |
| Discount rate | 10% |

Capital costs:

| | |
|---|---|
| - equipment CIF | $16052 |
| - installation | $ 8250 |
| - total | $24302 |

| | |
|---|---|
| Replacement costs, 2 pumps | $ 3500 |
| Total present worth of system | $27802 |

Recurrent costs:

| | |
|---|---|
| - maintenance | $  250 per year |
| - present worth | $ 1902 |

| | |
|---|---|
| Total present worth of life cycle costs | $29704 |
| Annual levelised cost | $ 3905 |

Unit Water Cost:

| | |
|---|---|
| - 100% use | $0.07 per m³ |
| -  50% use | $0.12 per m³ |
| -  25% use | $0.29 per m³ |

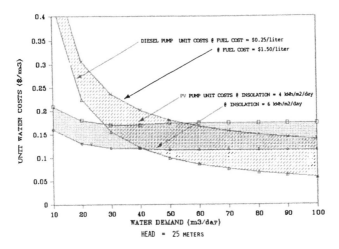

**Figure 3.4    Unit Water Costs**

Table 3.3  Data used for solar pumping cost comparison.

### PV PUMP SYSTEM CHARACTERISTICS

| | |
|---|---|
| Depth of water supply, head (m) = | 25 |
| Annual average daily water demand (m³/day) = | 20 |
| Annual max. daily water demand (m³/day) = | 30 |
| Insolation (kWh/m²/day) = | 5.0 |
| PV array peak power (kWp) = | 1.11 |
| PV pumping system capital cost ($/Wp) = | $12.00 |
| PV pumping system availability (%) = | 95.0% |
| PV array life (years) = | 20.0 |
| Pump life (years) = | 5.0 |
| Nominal discount rate (%) = | 10.0% |
| Inflation rate (%) = | 5.0% |
| | |
| PV NPV unit costs ($/m³) | $0.14 |

### DIESEL PUMP CHARACTERISTICS

| | |
|---|---|
| Depth of water supply, head (m) = | 25 |
| Annual average daily water demand (m³/day) = | 20 |
| Annual max. daily water demand (m³/day) = | 30 |
| Diesel generator power rating (kW) = | 3.0 |
| Average load factor (%) = | 45.4% |
| Diesel fuel cost ($/litre) = | $0.75 |
| Diesel gen-set pump capital cost ($/W) = | $1.59 |
| Diesel pump availability (%) = | 90.0% |
| Diesel gen-set life (years) = | 6.0 |
| Pump life (years) = | 5.0 |
| Nominal discount rate (%) = | 10.0% |
| Inflation rate (%) = | 5.0% |
| | |
| Diesel NPV unit costs ($/m³) | $0.26 |

them to be successfully integrated into the rural
communities that stand to benefit.

There are three main areas where institutional support is
particularly needed:

-    at the planning and procurement stage

-    for administering the operation

-    maintenance and spare parts.

At the planning stage, it is important to involve the local
community from the outset and encourage them to organise a
management committee. The local costs should be raised
locally, either in cash or in direct labour. Clearly,
experienced technical advice will be needed for the design
and procurement of suitable equipment.

The local organisation must then be assisted to organise
appropriate arrangements for distributing the water and
levying charges. In some cases, a village co-operative is
formed to administer these local aspects. It is helpful to
appoint a keeper or operator to watch over the system and he
will need to be given basic training in routine maintenance
and simple trouble-shooting.

With good design and the installation of the latest types of
system, system reliability should be good. However, there
will inevitably be faults arising from time to time which
cannot be fixed by the users. Established arrangements need
to exist for calling in technical support and for the
procurement of any necessary spare parts. Good
communications between the site and the source of support
are, of course, very desirable, but it has to be recognised
that in many rural areas the necessary infrastructure simply
is not available.

## 3.3 Photovoltaic Refrigerators for Rural Health Care

### 3.3.1 The Need for PV Refrigerators

In many developing countries, living conditions for the
majority of the rural population are poor and there is
widespread malnutrition combined with a high incidence of
disease. Infant mortality is particularly high in the rural
areas, where in some countries, as many as one third of the
children die before the age of two. Much of the disease
could be eliminated or controlled through mass immunisation,
but the practical problems involved are formidable. Most
countries are, however, making large efforts to improve the
quality of rural health care, including expansion of their
immunisation programmes.

Vaccines require refrigeration during transportation and storage to remain effective. It is important to maintain the vaccine 'cold chain' from the place of manufacture right through to the point of use. This imposes a major logistical problem because generally there are not reliable electricity supplies to operate conventional electric refrigerators in the rural areas where the clinics and health centres are located. Kerosene and bottled gas (LPG) powered refrigerators are available but their performance in many cases is not adequate and there are often problems in ensuring regular fuel supplies.

Solar photovoltaic refrigerators have the potential for better performance, lower running costs, greater reliability and longer working life than kerosene or LPG refrigerators, or diesel generators powering electric refrigerators. Recognising this potential, the World Health Organisation (WHO), the Centre for Disease Control (CDC), the US Agency for International Development (USAID), the European Community (EC) and other agencies have installed and evaluated many PV refrigerators throughout the developing world. At least 1500 PV medical refrigerators have been installed to date, mainly for testing and/or demonstration purposes. The stage has now been reached where a number of designs have been approved by the WHO, opening the way to wider implementation of this technology.

The use of PV refrigerators instead of kerosene or LPG units offers the following benefits:

i)   Elimination of fuel supply costs and delivery problems.

ii)  Reduced vaccine losses through improved refrigerator reliability, with associated reduced anxiety among medical personnel.

iii) Reduced maintenance workload for technicians and medical personnel, with associated cost and time savings.

iv)  Overall cost savings for the vaccine cold chain equipment.

v)   A more effective and sustainable immunisation programme, leading to reduced incidence of disease.

### 3.3.2 The Technology

Five alternative methods of solar powered refrigeration were surveyed by the WHO during 1980 (Ref. 3.13). These were: photovoltaic/vapour compression, photovoltaic/Peltier effect, solid absorption/zeolite, solid absorption/calcium chloride and liquid absorption/ammonia. Photovoltaic systems are the only type commercially available for vaccine storage (with the exception of one solid absorption ice making plant manufactured in Denmark). Of the photovoltaic systems, several manufacturers offer vapour compression systems in suitable forms for use in the vaccine cold chains.

A schematic diagram of a photovoltaic/vapour compression refrigerator is given in Figure 3.5. A PV array charges a battery via a charge regulator, to ensure that the battery is not overcharged. The battery powers a DC motor which is coupled directly to the compressor. The motor/compressor is usually manufactured as a hermetically sealed unit. The motor is of the electronically commutated brushless type. A second regulator is employed to ensure that the motor/compressor operates only within its rated power range and to prevent over-discharge of the battery. Freon refrigerant is used in the cooling cycle in the normal way, i.e. the cooling effect is achieved by the heat absorbed by the refrigerant as it evaporates in the evaporator. A thermostat switches the motor/compressor unit on and off as required. Some models have two compressors and thermostats, one each for the refrigerator compartment and the freezer compartment.

The insulation is normally of the expanded polyurethane type and double the usual thickness to reduce heat gain and thereby reduce the energy consumption and increase the time the refrigerator can maintain safe temperatures with no power. Most units are top opening, to reduce loss of cold air and often have a secondary hinged or removable cover under the main door. Figure 3.6 shows a typical example of a PV vaccine refrigerator.

### 3.3.3 WHO Specification

The vaccine capacities of solar refrigerators available or being developed vary widely, from 3.6 to 200 litres. The need for solar refrigerators is greatest at the peripheral health centres serving populations of 20,000 to 100,000. The quantity of packed vaccine needed to immunise fully 150 infants and their mothers is approximately 4 litres. There is, however, no general agreement yet on the best size for a PV vaccine refrigerator. Opinions differ on the quantity and volume of other biological products (e.g. blood) which might be stored in the health centre refrigerator and many people believe that a larger cabinet will have a wider market. It is also important for the system to have the capacity to freeze ice-packs which are used when transporting vaccine from the health centre for immunisation in the field. The ice production capacity is a significant load on the system and has a major influence on the PV array size and, hence, system cost.

In 1981, the WHO issued an outline specification for PV refrigerators which laid down minimum requirements covering

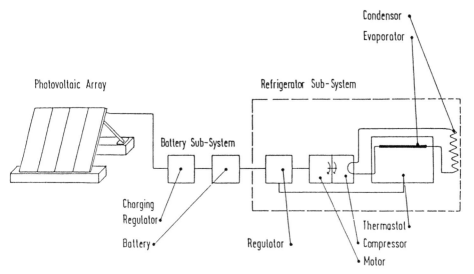

Figure 3.5   PV/Vapour Compression Refrigerator

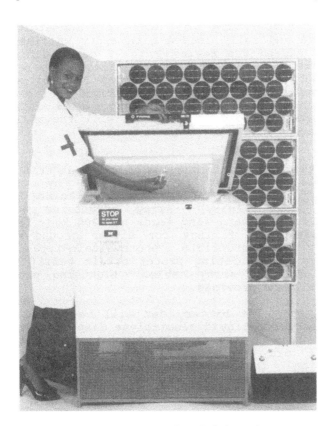

Figure 3.6  PV Vaccine Refrigerator

vaccine capacity, ice-making performance, refrigerator performance, hold-over time, battery maintenance interval, etc. The basic requirements have been modified in the light of field experience and the current WHO specification (Ref. 3.14) provides as follows:

1.  The design of the system will be sized to enable continuous operation of the refrigerator and freezer (loaded and included icepack freezing) during the lowest periods of insolation in the year. If other loads, such as lighting, are included in the system, they should operate from a separate battery set, not from the battery set which supplies the refrigerator.

2.  The design of the system will permit a minimum of five days continuous operation when the battery set is fully charged and the photovoltaic array is disconnected. During this time, the internal temperature of the refrigerator will remain within the range of +0 to +8 deg C when the constant external temperature is a minimum of +32 deg C.

3.  Refrigerator/Freezer: In continuous ambient temperatures of 20 deg C, 32 deg C and 43 deg C, the internal temperature of the refrigerator, when stabilised and fully loaded with empty vaccine vials, will not exceed the range +0 to +8 deg C. This range will be maintained when, in an ambient temperature of +22 deg C, the maximum recommended load of icepacks containing water at +22 deg C is placed in the freezer and frozen solid without adjustment of the thermostat. The recommended load of icepacks will freeze in less than 24 hours and will weigh at least 2 kg, without the material of the pack.

4.  Photovoltaic Array: Modules will meet the latest applicable specifications laid down by the Jet Propulsion Laboratory (USA) and the Joint Research Centre, Ispra (Italy). Array structures will be designed to withstand wind loads of +200 kg/m$^2$ and will be provided with fixings for either ground or roof mounting. Appropriate photovoltaic-type sealed connectors incorporating proper strain relief will be provided for the array cable. Lightning protection devices will be provided.

5.  Battery Set: The battery set will be sealed or low waterloss or non-liquid electrolyte deep discharge type (minimum 1000 cycles to 50% discharge). Automotive batteries are specifically unacceptable for this application. The batteries will be housed within the refrigerator/freezer cabinet, or in a cabinet separate from the refrigerator, but in either case, lockable.

No dry cell batteries shall be used to power instrument and controls.

6. Voltage Regulator: A voltage regulator will be provided, which meets the charge/temperature requirements of the selected battery and which cuts off the load when the battery has reached a state of charge which can be repeated to a minimum of 1000 cycles. Lightning protection will be provided. The load should be automatically reconnected when the system voltage recovers.

7. Instrumentation: A LED alarm will be installed to warn that power to the compressor has been cut by the regulator. An expanded scale voltmeter or a LED alarm will be installed to warn the user when the battery charge is in an unusually low state of charge to give adequate advance warning. The warning light of the minimum voltage limit should be clearly labelled "DO NOT FREEZE ICEPACKS" in the appropriate local language. If an external reading thermometer is provided for the refrigerator, it should be marked clearly in green between +0 and +8 deg C.

A thermostat or a defrost switch should be provided but no other power switches should be installed. Circuit breakers or cartridge fuse holders will be fitted with a polythene bag holding 10 spare fuses and special attention will be given to corrosion of fuse mountings.

8. Individual sea-crating of the components of each system should be provided whether or not containers are used to transport the systems. No package should be heavier than can be handled by hand in the country. Labels bearing handling instructions should be printed also in the appropriate local language.

9. Essential spare parts which may be needed during the first three years' operation should be assembled as a kit in appropriate quantities for central and regional storage in the country. A minimum list is as follows:

| Item | Quantity per 10 systems |
|---|---|
| 1. Photovoltaic modules | 2 |
| 2. Regulator components (sets) | 2 |
| 3. Battery sets | 2 |
| 4. Array cables | 1 |
| 5. Compressor, complete | 1 |
| 6. Spare compressor regulator cards | 3 |
| 7. Thermostat | 3 |

10 Manuals will be provided for the installation and use of each system.

The WHO guidelines also require the supplier to provide a
warranty for the replacement of any component which fails
due to defective design, materials or workmanship.    The
minimum period of the warranty is required to be 10 years
for the PV array, 5 years for the batteries and 2 years for
the remaining components.    The system supplier is also
required to provide technical support for maintenance and
repair operations in the country concerned for a period of
at least 2 years.   The supplier has to train an engineer to
assist with installation of each system and also train users
and repair technicians in each area.

### 3.3.4 Commercially Available Equipment

At least 20 companies now supply PV refrigerators for
vaccine storage.     To assist health authorities choose
appropriate equipment, the WHO Expanded Programme on
Immunisation (EPI) publishes Product Information Sheets
(Ref. 3.15).   Inclusion of a product on these sheets in
effect means that, based on the information and test results
available with WHO-EPI, the product is approved for use in
the vaccine 'cold chain'.

The PV refrigerators currently approved by WHO-EPI for use
in the vaccine cold chain are listed in Table 3.4. It should
be noted that there are many 12V DC-powered refrigerators
available, intended primarily for leisure applications in
boats and caravans.    These have been designed for low
capital cost, without consideraion of energy consumption or
internal temperature variation.   Although such systems can
be readily adapted for PV powering, they are not suitable
for vaccine storage.

### 3.3.5 Field Experience

The most significant work on testing and evaluating PV
refrigerators in the field has been carried out under the
WHO-EPI programme, involving field testing of over 50
systems for 12 suppliers in some 30 countries.    Other work
has been carried out by UNDP, UNICEF, European Development
Fund, AFME (France), GTZ (West Germany), ODA (UK), Oxfam and
other aid agencies.

Not all these projects have been consistently monitored, but
two major projects are in progress which should provide
substantial operating data in due course.    Both these
projects are funded by the European Development Fund, one
involving the installation of 100 systems in Zaire and the
other 20 systems in the South Pacific.    A summary of
experiences with photovoltaic refrigerators for medical use
is given in Ref. 3.16.

Table 3.4 PV Refrigerators Approved by WHO-EPI for Vaccine Storage.
(Data taken from EPI Technical Series No.1, The Cold Chain
Product Information Sheets, WHO, May 1988)

| System Supplier | Refrigerator/ Freezer Unit | Net Vaccine Capacity (litres) | |
|---|---|---|---|
| | | Refrigerator | Freezer |
| AEG (W. Germany) | Electrolux RCW 42 | 14 | 14 |
| AEG (W. Germany) | SET KT-180-24 | 56 | 0 |
| Ansaldo (Italy) | BP VR50 | 38 | 5 |
| BP Solar (UK) | BP VR50 | 38 | 5 |
| FNMA (Zaire) | FNMA 75 | 27 | 10 |
| Italsolar (Italy) | Electrolux RCW 42 | 14 | 14 |
| Italsolar (Italy) | BP VR50 | 38 | 5 |
| Leroy Somer (France) | Leroy Somer R50+IF50 | 16 | 16 |
| Noack Solar (Norway) | Polar Products RR2 | 80 | 20 |
| Noack Solar (Norway) | Marvel RTD4 | 80 | 9 |
| Noack Solar (Norway) | Electrolux RCW 42 | 14 | 14 |
| Polar Products (USA) | Polar Products RR2 | 80 | 20 |
| R & S (Netherlands) | Electrolux RCW 42 | 14 | 14 |
| RJM (Switzerland) | Electrolux RCW 42 | 14 | 14 |
| Solapak (UK) | Electrolux RCW 42 | 14 | 14 |
| Solarex (USA) | Polar Products RR2 | 80 | 20 |
| Solarex (USA) | Marvel RTD4 | 80 | 9 |
| Solarex (USA) | Leroy Somer R50+IF50 | 16 | 16 |
| T.I.S (USA) | Marvel RTD4 | 80 | 9 |

### 3.3.6 Conclusions

#### 3.3.6.1 Technical Aspects

Although some 1500 PV refrigerators have been installed to date, experience has shown that the technology has only recently matured. Laboratory testing by WHO and NASA-LeRC has prevented totally inadequate equipment being sent into the field, but some reliability problems are still being experienced. In particular, the sizing of the PV arrays and/or the batteries have been found to be inadequate for actual conditions, in particular in regions of high ambient temperature and poor insolation levels (e.g. Philippines, parts of India). Average availability has been around 80 to 85% for systems installed from 1981 to 1983 (Ref. 3.17). A summary of observed temperature control is given in Table 3.5.

Systems that have been installed more recently, particularly those from suppliers with previous experience, are being found to be more reliable. Some models are now showing 90 to 100% in-service time, with certain installations operating with 100% reliability for more than two years (Ref. 3.18). Some suppliers have withdrawn from the market (eg. SPC, WSR and Adler Barbour).

The problems of system sizing and load prediction remains a cause for concern. A recent evaluation of tenders for the supply of 23 PV refrigerators for islands in the South Pacific demonstrated that some tenderers proposed PV array sizes and/or battery capacities that would be grossly inadequate. Fortunately, many of the systems tendered were correctly designed. An easily applied method for prospective purchasers to check system sizing would be a significant help.

The ice making capability of most PV refrigerators commercially available is less than 2 kg/day. Some users have expressed the opinion that this is inadequate for many vaccine cold chains.

Battery maintenance has been a common problem with many systems. The possibility of developing a battery-less refrigerator making use of soft-start compressor motors and thermal storage instead of electrical storage should be given more attention as a development project.

The field trials have highlighted the need for a number of relatively minor improvements. These include the provision of door locks on some models and the relocation and/or redesign of some thermostats to reduce the possibility of unnecessary adjustments.

Table 3.5 Observed Temperature Control during WHO Field Trials on 7 Systems
(to June 1985)

| SYSTEM | PERCENT TIMES WITHIN TEMPERATURE BANDS | | | NO. DAYS DATA |
|---|---|---|---|---|
| | Correct 0-8 deg C | High >8 deg C | Low <0 deg C | |
| 1 | 89.92 | 10.08 | 0.00 | 295 |
| 2 | 86.22 | 0.00 | 13.50 | 711 |
| 3 | 71.72 | 0.17 | 28.11 | 471 |
| 4 | 82.25 | 11.45 | 6.29 | 3057 |
| 5 | 79.11 | 7.03 | 13.86 | 462 |
| 6 (*) | 73.86 | 0.00 | 1.59 | 440 |
| 7 | 62.56 | 1.37 | 36.07 | 219 |

Source : WHO

(*) Note: Data for this system does not total 100%. This is an
unresolved error in the original table.

3.3.6.2 Economic Aspects

Very little work has been undertaken on assessing the financial benefits of solar refrigerators using actual field data, but it is important to ensure that investment in PV refrigerators constitutes a sound use of development funds. The WHO-EPI is not an 'economic activity' in the normal sense and so it is not possible to carry out a cost-benefit analysis. The only meaningful analysis involved the comparison of costs of the various options and their likely influence on the achievement of the immunisation programme objectives. It is important to note in this regard that the fixed costs for any immunisation programme are generally large compared with the direct costs of vaccine refrigeration.

Kerosene refrigerators used in the vaccine cold chain have an initial capital cost of only $300 to $800, considerably less than for PV refrigerators. With transportation and installation, this may rise to $1500 installed compared with about $5000 for an installed PV system. The operation and maintenance costs of kerosene refrigerators are high however and their reliability is low, sometimes resulting in an avilability of only about 50%.

The results of a comparative cost analysis relating to an actual immunisation programme in the Gambia are presented in Ref. 3.19. based on data collected in 1984 and 1985.

It was concluded that the overhead cost per dose is reduced by $0.06 to $0.07 by using a PV refrigerator, due to the greater reliability. The refrigerator cost per dose is small compared with the overhead cost and is not significantly different between kerosene and PV. The overall cost per dose is cheaper for the PV refrigerator even where the PV capital cost is high.

It is important to note that periods when vaccinations cannot take place result in incompleted and hence ineffective courses of vaccinations. This effect is difficult to quantify but clearly favours the refrigerator with the higher reliability.

A life cycle costing comparative analysis is given in Ref. 3.16 which also concluded that the poor reliability of kerosene refrigerators makes photovoltaic refrigeration more economic in comparison. The results of this analysis are presented in Figure 3.7. The data assumed is given in Table 3.6.

3.3.6.3 Social and Institutional Aspects

Based on the reported experience of PV refrigeration projects to date, there is no doubt that the systems are

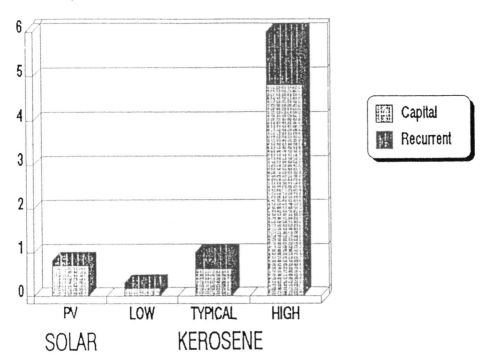

Figure 3.7    Life Cycle/PV Refrigeration

Table 3.6   Data assumed in unit cost comparison presented in Figure 3.5

| REFRIGERATOR TYPE<br>PARAMETER | PHOTO-VOLTAIC | KEROSENE<br>Low Case | Typical | High Case |
|---|---|---|---|---|
| NET VACCINE CAPACITY (Litres) | 100 | 100 | 100 | 100 |
| INITIAL CAPITAL COST ($) | 4500 | 300 | 500 | 1000 |
| CIF AND INSTALLATION ($) | 1500 | 500 | 800 | 1000 |
| FUEL COSTS ($/day) | - | 0.25 | 0.70 | 3.00 |
| MAINTENANCE COSTS ($/year) | 150 | 50 | 100 | 150 |
| LIFETIME (years) | 15 | 10 | 5 | 2 |
| AVAILABILITY (% time in service) | 95 | 80 | 50 | 20 |

widely acceptable to the users. The main need is to ensure a considered approach is taken to project implementation, in particular with respect to:

- project design (selection of systems and sites)
- selection of local implementing agencies
- user training (operation and maintenance and trouble shooting)
- technical support centres serving each region using PV refrigerators
- communications with technical support centre.

The recommendations of the WHO-EPI with regard to procurement of approved equipment and the contractual arrangements for warranties and training of operating and maintenance staff should be followed by all authorities wishing to install PV refrigerators for the vaccine cold chain.

WHO-EPI have now commissioned the preparation of installation, user and repair technician handbooks for photovoltaic refrigerators for use in the vaccine cold chain. In addition, regional technician training courses have been run by WHO-EPI. Both of these initiatives should assist with the successful introduction of solar refrigeration into vaccine cold chains.

There is a continuing need to gather data on system performance and therefore efforts should be made to provide the necessary instrumentation and organisation required to monitor the systems in the field. The information, both quantitative and qualitative, should be passed on to the EPI co-ordination office in Geneva. The following minimum information should be recorded on a daily basis:

- maximum and minimum internal temperatures of the refrigerator and freezer
- ambient air temperature
- solar irradiation
- system usage (kg of ice removed, vaccine removed)
- details of breakdowns or component faults.

For system optimisation studies, the electrical energy delivered by the PV array and the energy consumed by the motor/compressor unit is also required. The preferred method of data collection is to use data loggers supplemented by a log book or pro-forma sheets for noting details of system use and reliability.

## 3.4 Lighting

### 3.4.1 Alternative Lighting Techniques

Lighting is a steadily growing need in the rural areas of developing countries, not only because the population is increasing but also because more people want to be active in the evening. School children need to study and there are new work and leisure opportunities for adults. An important need is for lighting for small commercial enterprises in the streets, such as food stalls, shops and recreational activities. In addition to these residential and commercial needs, there is an associated need for lighting for streets and public open spaces.

In areas where there is no electricity supply, lighting for domestic and commercial applications is usually provided by kerosene lamps or candles. In general, lighting from these sources is of poor quality, expensive and a fire hazard. The best light using kerosene comes from a pressure device (Coleman type), but these are expensive. The more commom wick devices (hurricane lamps) produce less than 15% of the light of a 20 W fluorescent tube. Due to the high price of kerosene in remote areas, a household may have to spend the equivalent of over $200 a year to operate two kerosene lamps.

Photovoltaic lighting systems would be an attractive alternative to kerosene lamps and candles throughout the areas where it is likely to be many years before regular electricity supplies become available. The key considerations are comparative quality, reliability and cost.

### 3.4.2 Technical Requirements

PV lighting systems have become readily available over the last five years, with manufacturers offering two basic types of unit, one for area lighting, the other for domestic applications.

Area lighting units may be used for street lighting, public open spaces and security lighting. These systems consist of PV array; battery; simple voltage regulator; timing and/or photosensitive switch controls; and an efficient fluorescent or low-pressure sodium or mercury vapour lamp. Several manufacturers offer complete self-contained units including poles with mountings for the lamp and the PV modules and a weather-proof container for the battery and controls. As the pole represents a significant proportion of the total cost, some manufacturers supply only the PV array, lamp, battery and controls, to allow the purchaser to provide the pole from local sources.

Domestic lighting units typically require only one or two PV
modules for charging a battery which supplies from one to
four fluorescent tubes, from 20 W to 40 W depending on the
application.  Some systems are portable, with a lantern unit
incorporating a rechargeable battery.  Larger systems can be
obtained, capable of supplying other end uses such as
refrigerators, radios and televisions, but it is more
appropriate to consider these systems in the next section
under the general heading of rural electrification.

Fluorescent lamps are commonly used for both area lighting
and domestic lighting systems.  Fluorescent (or gas vapour)
lamps offer high efficiency, long life and a high
reliability.  They require a 'ballast' and a 'starter' which
gives a high frequency impulse for starting, followed by
much lower power and frequency for normal running.  Standard
AC fluorescent units may be converted for DC powering (and
therefore suitable for PV systems) by changing the ballast
and starter components, a relatively simple task.

### 3.4.3 Field Experience

Several thousand PV lighting systems are in use in
developing countries.  Experience is particularly extensive
throughout the South Pacific and more specifically in Papua
New Guinea, Fiji and French Polynesia (Refs. 3.20-3.24).

There are privately-funded schemes in some countries to
enable the benefits of PV lighting systems to be accessible
to relatively poor people.  In the Dominican Republic, for
example, a US-based organisation distributes PV lighting
systems to villagers in the northern part of the country,
with loan finance repaid over two to five years.

### 3.4.4 Conclusions

### 3.4.4.1 Technical Aspects

PV lighting systems covering a wide range of sizes and types
are widely available as standard products.  The components
required for a typical domestic lighting system are listed
in Table 3.7.  Such a system would provide up to 200 Wh/day
of useful energy for lighting given a solar input of 6.0
$kWh/m^2$.  The 20W lamp could be used for 3 to 5 hours every
night for general activities and the 7W lamp could be used
for 8 to 12 hours for security.  The battery provides about
three days storage.

Many of the smaller systems for domestic use are portable,
which makes them particularly suitable for use in place of
kerosene lamps.  The introduction of long-life rechargeable
Ni-Cd batteries is an interesting development in this
regard.

The battery charge controllers used for some early designs of PV lighting systems were found to be unreliable, but now fully tropicalised units are supplied which have proved very reliable in practice. The reliability and efficiency of the ballasts used in commercial fluorescent lamp units have been found to be variable. Careful selection of this component is therefore essential, particularly as it accounts for up to 75% of the cost of the lamp. The lifetime of the DC ballasts used for some 60 DC fluorescent fixtures tested as part of the NASA-LeRC PV medical refrigerator programme was found to be less than five years. Further experience of the lifetime of these components under field conditions is needed.

Small PV systems for domestic lighting are widely available and several thousands have been installed, particularly in the South Pacific, French Polynesia and China. They are simple to operate and reliable, now that the earlier problems with battery charge regulators have been solved. PV powered fluorescent tubes provide a much higher quality of light compared with candles or kerosene wick or pressure lamps. The efficiency of the DC ballasts used for fluorescent tube lamps is a key technical factor in the design of systems.

PV powered street lights and security lights are also available from several manufacturers. These units generally use low pressure sodium vapour or high pressure mercury vapour lamps. Some problems connected with the need to adapt standard AC units for DC operation have been experienced and there is need for further development of suitable ballasts. Specially designed AC systems for high-mast lighting have also been demonstrated. The key factor here is the performance of the inverter.

### 3.4.4.2 Economic Aspects

Lighting is not a directly economic activity and therefore a cost/benefit analysis is not possible for this application. A number of cost comparisons for alternative lighting methods, including the projects detailed earlier, indicate that PV lighting systems offer the cheapest solution for lighting in villages where grid electricity is not available.

It should be noted that the light output of a kerosene pressure lamp is about 200 lumens. That for a kerosene hurricane lamp is about 80-100 lumens, whereas the light from a 20W fluorescent tube is about 1000 lumens. Also the life of the PV modules (the most expensive part of the PV systems) is at least 15 years, much longer than the life of kerosene lamps.

Based on an analysis over five years with 10% interest rate,
PV lighting systems are found to be cost competitive with
kerosene lamps in areas where the cost of kerosene is
$0.75/litre or more.   Thus, given suitable finance, users
would be able to save money and improve the quality of
domestic lighting by installing PV lighting.   Several
organisations are operating successful financing schemes
which provide a capital grant plus a loan for the balance
repaid over five years at 9 or 10% interest.

Provided suitable means are available to finance the initial
cost with repayments over say 5 years, PV systems should be
widely attractive on both cost and performance grounds to
potential users in the villages.

### 3.4.4.3 Social and Institutional Aspects

For the successful introduction of PV lighting systems, the
potential users need first to be convinced of their
technical performance and reliability.   This requires
demonstration systems to be available, possibly as mobile
units to be taken from district to district.

There are then two major institutional requirements:

a)    The provision of finance, probably through the
      provision of a subsidy and low-cost loan, repayable
      over two to five years;

b)    The provision of technical support, in particular for
      the supply of spare fluorescent tubes and ballast and
      batteries.

In many cases, it would be preferable for the implementing
organisation to establish local co-operatives, who can
arrange for the administration of funds and the provision of
technical supports.   Training for key personnel is needed.
The local co-operative would need to be able to refer major
problems to a central resource centre.

Although in the short and medium terms, most countries would
need to import the PV modules, most other components could
be locally manufactured and assembled, thereby greatly
reducing the foreign exchange requirements whilst at the
same time building up local technical skills.

### 3.5 Rural Electrification

### 3.5.1 Conventional Approaches

There are two main techniques used for rural electrification
at present: either the main electricity grid is extended to
cover the selected area, or a diesel generating station is
established to serve a small network not connected to the

grid. Both approaches have associated economic and technical problems. Extending the grid over long distances is expensive and, in the initial years at least, the loads are often small, resulting in load factors and stability problems for the system as a whole. Maintaining long transmission lines over difficult terrain presents difficulties for the utility.

Diesel generators require regular supplies of fuel, which often presents problems for remote areas, particularly at certain seasons of the year when roads are practically impassable. Moreover, any national fuel shortages are likely to impinge more severely on the rural areas. In addition to the provision of fuel, it is often the case that the operating utility finds difficulty in retaining competent operating and maintenance staff, since as a para-statal organisation they are unable to offer competitive salaries. Even given the necessary staff, it may be practically impossible to obtain the necessary spare parts to keep the engines running.

Rural electrification is regarded as a development priority in most developing countries, for social and economic reasons. Although vast sums are expended each year on rural electrification project, it will nevertheless be many years before villages that are a long way from the main electricity grid lines or from the nearest all-weather road will benefit from a reasonably reliable and affordable electricity supply.

There is no doubt that new and renewable energy sources would be preferred for rural electrification, if systems were available at acceptable costs and with proven reliability. In some locations, wind generators or micro-hydro systems may be feasible, but in general photovoltaic systems would be favoured if the costs were right, since they involve no mechanical moving parts and require only simple maintenance.

## 3.5.2 Design and Load Estimation

For many households in developing countries, the main use for electricity would be for lighting. Many thousands of small PV systems dedicated for lighting applications are already in use worldwide, as discussed in the previous section. For many households, however, electricity would be useful for many other applications. Some of these could also be covered by small dedicated stand-alone systems, such as systems for water pumping or rice milling as discussed elsewhere in this chapter, but in general it would be preferable for rural electrification schemes to be non-specific, with the electrical power available for any small-scale application.

The great majority of users would be domestic households. Based on data presented in Refs. 3.25 and 3.26, for design purposes the peak demand and daily energy consumption for a typical household would be as shown in Table 3.8. Although, in general, it may take several years to build up to the power and energy levels shown, as most households could not afford all the appliances involved at once, it is prudent to plan a rural electrification scheme for the situation likely to arise within a few years.

There will of course be a number of connections which exceed this load estimate, as well as many connections which are much less. There are differences between countries, depending on standard of living and social patterns (e.g. number of persons living in one household). In addition to estimating domestic load, it is necessary to calculate specific loads for commercial, institutional and industrial applications.

### 3.5.3 Household versus Central Systems

To date, most PV demonstration projects for rural electrification have consisted of either small packaged systems for lighting or other specific end uses, or larger centralised projects serving a whole village. For a demonstration project, it is administratively easier to install a single central unit to serve a whole village. There is no doubt that the larger the system, the greater the publicity. There are, however a number of important advantages to be gained by installing individual household systems rather than a central PV plant. The main reasons for preferring household systems (from Ref. 3.25) are as follows:

a)    The PV arrays can be roof mounted, out of reach of livestock and people and not taking up valuable land;

b)    A distribution system is not required to take the power to each house, an expensive item if the houses are widely spaced;

c)    No metering system is needed, thereby avoiding the associated adminstrative costs for meter readers and billing computations;

d)    With no distribution system, the problems associated with unauthorised connections and theft of electricity are avoided;

e)    A centralised system may soon prove unreliable, due to its complexity and to possible overloading aggravated by unauthorised connections; failure of the system affects everyone, whereas failure of a household system affects only one consumer.

Table 3.7 Typical PV lighting system for domestic use

| Specification | Life (years) | Price ($) |
|---|---|---|
| 1 PV Module - 40 Wp | 15 | 250-300 |
| Battery - 12V/105Ah | 4 | 50-80 |
| Fluorescent lamps | 2 | 50-70 |
| Battery charging controller | 5 | 50-100 |
| | Total | 400-500 |

Table 3.8 Typical design for electrical load for rural household.

| LOAD | POWER (W) | DURATION (hrs/day) | ENERGY (Wh/day) |
|---|---|---|---|
| Lights 3 x 20 = | 60 | 5 | 300 |
| Fans  2 x 60 = | 120 | 6 | 720 |
| TV | 60 | 4 | 240 |
| Iron | 200 | 1 | 200 |
| Miscellaneous | 60 | 1.5 | 90 |
| Total | 500 | | 1550 |

Although a centralised system may appear a little cheaper in
initial cost, even after allowing for the cost of the
distribution system, this is heavily outweighed by the
reduced operational problems and administrative costs
associated with separate household systems.

### 3.5.4 System Features

If PV generators are used for rural electrification, whether
through household systems or centralised plants, the DC
output would normally need to be converted to AC at the
national standard voltage and frequency. Although most of
the loads such as lighting and domestic appliances could be
designed or adapted for DC operation, this would be
unpopular with consumers, who would wish to have the freedom
to buy cheaper standard AC products from the normal
suppliers in the markets. DC equipment would introduce too
many complications. A typical PV system for rural
electrification by means of a central plant is shown
schematically in Fig. 3.8. The schematic arrangement for a
household system is similar, as shown in Fig. 3.9.

Central plants would need to have battery storage for about
three to five days supply, depending on the Loss of Load
Probability (LOLP) level considered to be appropriate. The
battery storage for household systems could perhaps be less,
say two to three days supply, since energy management
procedures could more readily be introduced by the user when
an alarm on the system indicates that battery charge is low.

An energy meter would not be required for a household
system, since it would be most appropriate to charge the
consumer a constant monthly or quarterly rent for the
system. Although the consumer would need to understand the
load limitations of his sytem, there would be no particular
advantage in trying to economise on the use of electricity,
at least all the while the battery state-of-charge was
within its normal operating range.

### 3.5.5 Commercial Equipment

There are no standard systems for central PV plants,
although a number of demonstration plants have been built
and all the major PV manufacturers are able to offer a
design service for such plants. Several of the PV pilot and
demonstration plants sponsored by the European Community are
for rural electrification (e.g. Aghia Roumeli, Crete,
Greece, 50 kWp and Kaw, French Guyana, 35 kWp). In addition
to the central plants, there is also an EC-supported
demonstration project currently in progress which provides
for the installation of small stand-alone PV generators for
some 40 houses in the south of France and in Corsica

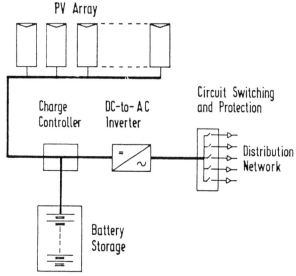

Figure 3.8    PV System for Rural Electrification

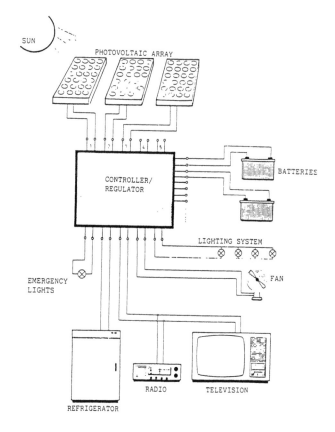

Figure 3.9    Schematic Arrangements for Household System

involving three standardised designs: 400 Wp, 800 Wp and 1200 Wp.

A number of village electrification projects involving photovoltaic systems have been undertaken in developing countries. Two villages in Indonesia have been equipped with central PV plants and there are similar schemes in Tunisia, French Guyana and Senegal. There are also a number of villages on Greek islands in the Mediterranean equipped with central generators.

In addition to these and other central plants, the alternative disaggregated approach using stand-alone systems for households and specific end uses has been followed in French Polynesia, Gabon and elsewhere.

### 3.5.6 Conclusions

#### 3.5.6.1 Technical Aspects

Rural electrification differs from lighting in that it is a more systematic approach to providing an electricity supply to a village or district. The electricity is then available for a wide range of domestic, commercial and agro-industrial applications. A number of relatively large central PV demonstration plants have been built to electrify a complete village, but there are a number of technical and institutional problems with this approach. Central systems of this type are vulnerable to failure due to component faults or over-loading and thus need a high level of supervision.

A more viable approach to rural electrification is to equip each household with its own stand-alone system. A number of standard sizes would be required to suit the size of the household and the nature of the loads likely to be imposed. Larger systems would be needed for commercial, institutional and industrial premises.

The electricity produced for general applications, such as lighting, fans, refrigerators, radios and televisions, needs to be AC, at the national voltage and frequency. This means that the PV systems have to incorporate inverters, which have not proved particularly efficient or reliable in the past. However, several manufacturers are now introducing new inverter types which offer much improved performance. At least one type switches itself into stand-by mode when no loads are connected, thereby greatly reducing the internal system losses. Initial field experience with these inverters is encouraging.

There is limited experience available to date on the performance of larger centralised PV plants for village electrification. Some systems have worked well, others have experienced problems with inverters and other components. No standardised designs exist as yet and care must be taken at the design stage to ensure that only components with proven reliability are used in systems that are to be installed at remote sites.

There is however a considerable volume of experience with smaller household systems, with many hundreds of installations in French Polynesia and elsewhere. Standardised designs have been developed and performance is generally reported to be very satisfactory.

### 3.5.6.2 Economic Aspects

There is little economic data on the cost installing, operating and maintaining centralised PV plants for village electrification. There is however extensive information on the economics of household systems. The current (1988) installed cost of a small (less than 500 Wp) stand-alone PV generator (including batteries and inverter) is about $15/Wp if bought as a one-off item. If however, standardised systems are bought in large quantities and integrated with the building, it is possible to get installed system costs down to around $5/Wp. Based on the low price scenario for PV modules and system presented by Starr, a cost projection for standard 360 Wp PV systems suitable for stand-alone household AC generators can be constructed, as shown in Table 3.9. The price per peak Watt is shown as falling from a current level of $15 to $5 by the year 2000.

It is now relatively straight forward to calculate the average unit cost of electricity produced by the PV generator for the range of system costs shown in Table 3.10. Annual maintenance expenses are assumed to be 1.5% of the initial capital cost. Administrative expenses are assumed to be $10 per system per year, given that there are a large number of systems within each administrative area. The average electricity cost is based on a 20 year period of analysis and a 10% discount rate. The average solar input is assumed to be about 5 kWh/m$^2$ per day. Each system is assumed to give a useful energy output of 540 kWh/year. The resulting average energy costs are as follows;

| Installed system cost ($/Wp) | 15 | 10 | 7 | 5 |
|---|---|---|---|---|
| Average energy cost ($/kWh) | 1.34 | 0.90 | 0.64 | 0.46 |

These unit energy costs may then be compared with the real (unsubsidised) cost of electricity supplied by alternative means, such as grid extension or by diesel generators. Grid extension costs depend on two main factors; the distance the feeder line has to be extended to serve the new area; and the number of connections served by the new feeder. Based on typical cost data, the unit energy costs for grid extension schemes are as set out in Table 3.10 for three

Table 3.9 Cost Projection for Household PV System (from Ref. 3.26)

| YEAR:<br>COMPONENT | 1985 | 1990 | 1995 | 2000 |
|---|---|---|---|---|
| Modules (360 Wp) | 1800 | 1080 | 810 | 540 |
| Battery (299Ah/48V) | 1600 | 1200 | 800 | 440 |
| Inverter/Regulator | 1000 | 600 | 300 | 250 |
| Building related costs<br>transport, installation etc | 300 | 300 | 300 | 300 |
| Supplier's mark-up | 700 | 470 | 330 | 270 |
| | ---- | ---- | ---- | ---- |
| Total installed system<br>price | 5400 | 3650 | 2540 | 1800 |
| Price per peak Watt | 15 | 10 | 7 | 5 |

Note: all costs in 1985 US dollars

Table 3.10 Unit Energy Cost for Grid Extension (from Ref. 3.26)

| | Feeder<br>length<br>(km) | No. of connections per feeder | | | | |
|---|---|---|---|---|---|---|
| | | 100 | 200 | 500 | 1000 | 2000 |
| Unit energy cost | 10 | 1.00 | 0.67 | 0.47 | 0.40 | 0.37 |
| in $/kWh | 20 | 1.69 | 0.97 | 0.61 | 0.47 | 0.40 |
| | 30 | 2.35 | 1.34 | 0.74 | 0.54 | 0.44 |

values of feeder line length and five values of the number of connections per feeder.

The unit energy costs for the PV and grid extension approaches are compared in Figure 3.10. Grid extension is the cheaper alternative for the near future except for the occasions when it is necessary to provide electricity to a limited number of connections involving long feeder lines, it is likely that PV systems will be cheaper than the grid if the grid is more than a few hundred metres away.

Diesel generation costs are summarised in Table 3.11 based on data published in Ref. 3.27. The unit energy cost for a system supplying 540 kWh/year (about 1.5 kWh/day) ranges from $1.00/kWh for the low cost case to over $2.50/kWh for the high cost case. Larger diesel generators serving whole villages typically give costs from $0.60/kWh to $1.50/kWh, given reasonable maintenance and no major interruptions in the supply of fuel and spare parts.

From these data, it is clear that PV systems are cost-competitive today with small diesel generator systems. They can also be competitive with larger diesel generator systems in places where fuel supplies and maintenance present major difficulties. PV systems will become increasingly competitive with grid extension schemes as the cost of PV modules and other components continues to fall with improved technology and larger volume production.

### 3.5.6.3 Social and Institutional Aspects

Whatever method is adopted for rural electrification, costs are bound to be high. Many developing countries have established rural electrification boards (REB's) to undertake the necessary planning, implementation and operation of rural electrification schemes. The REB's normally require substantial external funding, since it is widely recognised that rural electrification projects can rarely be self-financing. The selling price of electricity to rural consumers has to be kept at a low level, comparable to that obtained in urban areas, since (a) most rural consumers are very poor and (b) any significant disparities would generate strong political and social pressures. The cash flow for a typical REB with an expanding programme is thus bound to be poor, particularly as it often takes several years for the loads and associated revenues to build up.

Rural electrification schemes therefore have to be economically appraised and justified on broader grounds than simply costs and revenue from the electricity system alone. The value of the social benefits that accrue to the community as a whole, through the raising of living standards, improvements to land and labour productivity and

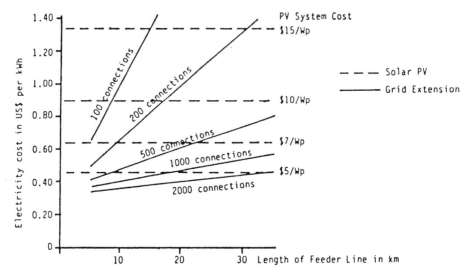

Figure 3.10    Unit Energy Costs for PV and Grid Extension

Table 3.11 Diesel Generator Costs

|  | Low Cost | High Cost |
|---|---|---|
| Size of diesel (kW) | 2.5 | 2.5 |
| Life of system (years) | 7 | 7 |
| Capital cost ($) | 1000 | 3000 |
| Fuel cost ($/litre) | 0.40 | 0.80 |
| Overall efficiency (%) | 20 | 15 |
| Operating O&M ($/year) | 200 | 400 |
| Unit energy cost in kWh for: | | |
| 0.5 kWh/day | 2.30 | 6.00 |
| 1.0 kWh/day | 1.50 | 3.30 |
| 2.0 kWh/day | 0.80 | 2.10 |

the generation of new employment opportunities, all need to be assessed. Thus a rural electrification scheme may be economically justified, even though the electricity has to be supplied at a loss by the utility. This normally requires a substantial financial subsidy to be made to the REB by the government. The true cost of supplying electricity to rural areas may often be more than $1.00/kWh, whereas the price to consumers may only be $0.08/kWh or less.

Sometimes the full value of the subsidies involved is hidden in the accounts of the electricity supply utility, especially in countries where there is no separate REB. This situation can arise when rural electrification costs are not distinguished from overall costs and a uniform tariff is applied throughout the country, for both urban and rural areas.

Thus, when comparing the costs of alternative techniques for rural electrification, whether grid extension, isolated diesels or photovoltaics, it is important to ensure that similar assumptions are made regarding the value of subsidies that are made available.

If it is decided to proceed with a rural electrification scheme based on PV systems, it is vital to ensure that the users of the new technology are supported by adequate arrangements for technical assistance and the supply of spare parts. This will call for good information programmes to help users understand how to get the best out of their systems and to identify faults. Technicians will need to be trained to install systems, instruct users in operation and routine maintenance, and carry out more extensive maintenance operations when necessary.

Any scheme for rural electrification using PV systems will need to be planned to provide for public information and other forms of institutional support. This would probably be best organised within the existing rural electrification board, to cover all technical, administrative and financial aspects.

### 3.6 Other Applications

### 3.6.1 Agricultural Applications

Photovoltaic systems can be used for a number of applications related to agriculture. Water pumping for irrigation is the major use, as discussed earlier in this chapter. Other applications include agricultural product processing (e.g. grain milling), milking machinery, cattle fencing, refrigerated storage of perishables and ice production for fish preservation.

All these systems are similar in that they employ a PV array to charge a battery which then supplies the end use, either direct for DC applications or through an inverter for AC applications. Because of the high cost of the PV array and batteries, it is necessary to optimise the design of the system as a whole. It is not normally advisable to use standard AC or DC appliances, since these, although relatively inexpensive, often have poor efficiency.

Most PV manufacturers offer standard systems for battery charging, but specialised advice is generally needed for the selection of appropriate appliances, such as DC motors for grain mills, or inverters for AC systems. Several manufacturers offer PV powered cattle fencing systems, consisting of a small PV module for charging a battery through a regulator and a standard high voltage pulse generator. These systems are beginning to find wide acceptance in areas where it is difficult to arrange for cattle fence batteries to be recharged regularly. These electric fences can be used not only to keep domestic livestock within required limits, but also to keep wild animals out of areas where they could damage crops.

There are not many examples of dedicated systems for agricultural product processing, as usually the relatively small electric motors are powered from a larger system supplying a number of end uses. Besides water pumping for irrigation, grain milling is probably the most important agricultural use. Although not many PV powered grain milling systems have been built, there is no reason to doubt the technical feasibility of this application. A PV powered grain mill has been operating successfully for eight years in Tangaye, Burkina Faso. It forms part of a 3.6kWp PV system which also includes a water pump. The grain mill has been modified twice since the time of installation in 1979 to obtain the required degree of fineness and consistency, but the overall availability has been reported as over 90%. Such applications are likely to be economic in comparison with diesel powered systems in remote areas, provided the demand is reasonably uniform throughout the year.

Although no installations for milking machines are known in developing countries, a large PV generator for a milking parlour has been built as a research project in Ireland. The 65 kWp grid connected system built on Fota Island in 1982 provides electricity for milking machines and milk processing equipment on a large dairy farm. The performance is being closely monitored by the University of Cork and the reliability is reported to be very high (Ref. 3.28).

Another research project in Europe is the PV powered refrigerated cold store for agricultural produce built on Giglio Island, Italy, which started operation in 1984. A 45 kWp PV array provides power for an ozoniser (for water

disinfection) and a cold store of about 275 m$^3$. A
particular feature of the compressor used in the
refrigeration plant is the control system, which selects the
number of cylinders to be loaded to match the power
available from the PV array (Ref. 3.29).

These large PV systems for agricultural applications are
still at the development phase. The smaller systems, for
powering grain mills, cattle fencing, irrigation pumps and
so on, are much simpler in concept and are technically
developed. They are most likely to be cost effective in
rural areas of developing countries where the following
conditions obtain:

a)      There is no electricity supply from the grid and the
        costs and practical difficulties of running diesel
        engines are high;

b)      The solar input is reasonably good throughout the year,
        with an average of at least 4.5 kWh/m$^2$ per day;

c)      There is a need for the application for the major part
        of the year, as the high capital cost requires a high
        utilisation factor to achieve low unit costs.

Two important factors affecting the success of any attempt
to introduce new technologies into the rural sectors are (i)
the institutional arrangements for technical training and
support and (ii) the local management of the system. It is
significant that the success of the PV grain mill in Tangaye
is largely attributed to the way existing practices were
adapted to form a cooperative to manage the installation.
This gave a sense of communal ownership, encouraging
interest and concern for the success of the project.

### 3.6.2 Water Treatment Systems

Several PV manufacturers offer complete PV powered water
treatment systems. In some systems, the water is first
filtered and then given a prophylactic chlorine dose,
usually in the form of sodium hypochlorite, before being
transferred to storage and thence to consumers. In view of
the difficulties in ensuring the supply of hypochlorite and
its correct use when available, some system designers are
concentrating on chemical-free water treatment processes,
which involve slow sand filtration for sterilisation. This
approach removes all harmful bacteria, but care has to be
taken to avoid subsequent contamination.

In these two approaches, the PV power is used only for
pumping the water, first from the source, such as a borehole
or river, and then through the various filtration stages to
the final storage reservoir. Another type of water

treatment system being developed incorporates a PV powered UV light for sterilisation.

These complete water treatment packages using PV power are likely to be of particular interest for water supply projects in remote areas where the existing water sources are known to be heavily polluted.  Two British designed systems are known to be operating in Nigeria. The use of a solar pump combined with a slow sand filtration system has proven highly successful in Indonesia, where there is an abundance of surface water which is polluted (Ref. 3.30).

The emphasis for design of these systems is on reliability and minimum requirements for chemicals and other consumables.  Several approaches have been demonstrated, including systems involving slow sand filtration, filtration followed by a prophylactic dose of sodium hypochlorite and filtration followed by UV sterilisation.  Experience is too limited for any general conclusions to be made regarding the best approach.  Costs are likely to be competitive with any alternatives involving diesel generators.

### 3.6.3 Telecommunications

Telecommunication systems in developing countries have traditionally been powered by grid electricity or stand-alone diesel generators.  Battery banks are usually provided for security of supply in the event of power interruptions. Problems of unreliable supply, variable quality (voltage spikes, low voltage) and high cost of operation and maintenance cause constant problems for the system operators. The quality of communications frequently suffers as a result.

Photovoltaic generators are particularly suited for telecommunication systems, since they can provide the relatively small amounts of power required at remote transmission/reception sites reliably and with little or no maintenance.  PV generators are widely used for this application and many hundreds of systems are operating worldwide, including places where the average solar input is as little as 2.5 kWh/m$^2$ per day.  In fact, photovoltaic systems have had more commercial success for telecommunications applications than any other remote power application.

There are three main types of telecommunication system which can be PV powered:

i)   Two-way radios, including radio telephones;

ii)  Radio and television secondary (infill) transmitters;

iii) Telephone systems, including exchanges, repeater stations and satellite ground stations (see Fig. 3.11).

In addition, television sets can have a dedicated PV power supply and educational TV is an application which has found wide use in certain West African countries for use in village schools.

In each case, the PV system is primarily required for battery charging. For the larger systems, a stand-by diesel generator may be provided, with controls for automatic start-up if the battery voltage falls to a pre-set low level. The hybrid arrangement is optimised for the least cost configuration.

PV systems are likely to be found cost effective for sites where grid electricity is not available for loads up to about 2.5 kWh/day for most locations. In some circumstances, PV/diesel hybrid systems may be found cost effective for loads up to 10 kWh/day.

### 3.6.4 Cathodic Protection

Another application where PV systems have found substantial commercial markets is for cathodic protection of steel pipelines and other steel structures. There are many examples of such systems used for oil and gas pipelines in the Middle East. Cathodic protection by the impressed current method involves maintaining the steel structure at a negative potential with respect to the surrounding soil or atmosphere. PV systems are particularly well suited for this application, since they provide the necessary DC power without the need for transmission lines, transformers and rectifiers as required for grid powered systems. An example is shown in Figure 3.12.

### 3.6.5 Unusual Applications

Photovoltaic systems should be considered wherever there is a requirement of a small amount of power in remote or inaccessible locations. For example, aircraft warning lights on tall structures or on hilltops, or navigation lights marking out the channel into a harbour. One manufacturer in Europe has developed a mobile orthopaedic clinic powered by a PV generator to enable a full range of equipment to be used in areas which have no electricity supply.

The eleven large steel lattice towers carrying a 230 kV double circuit transmission line 15 km long across the Jumuna River in Bangladesh are each fitted with five aircraft warning lights powered by a 700 Wp PV array (Ref. 3.31). Battery storage is sufficient for 10 days supply. Other methods of supplying the lights considered but rejected were diesel generators (access problems for fuel

Figure 3.11   Telephone Systems

Figure 3.12   Cathodic Protection

supplies), induction from the phase conductors (no power during times of outage) and earth wires from the local network (unreliable).

There are now many navigation buoys and lighthouses supplied by PV generators. The cost savings can be substantial, as servicing the batteries in navigation buoys or the diesel generators associated with lighthouses is expensive and sometimes hazardous. There are five lighthouse systems ranging from 2.6 kWp to 18.2 kWp operating or under construction in the Mediterranean (Greece, Italy, France) plus several hundred PV powered navigation lights for harbour entrances and buoys along the southern coast of France. There are similar developments in many other countries.

PV systems can also provide power conveniently and economically for remote metering installations, such as river gauging station, meteorological stations and ground water level monitoring systems. The records can either be transmitted by radio to the control centre or stored on tape or disk for later collection. Some PV installations in France use satellites (Argos and Meteosat systems) for data transfer.

## 3.7 Conclusions on Photovoltaics for Developing Countries - Summary of Experience

### 3.7.1 General

Several thousand photovoltaic systems have been installed in developing countries over the past ten years, the great majority over the past five years. The size of these systems ranges from a few Watts to over 30 kWp, for applications as diverse as water pumping, vaccine refrigeration, domestic lighting, cattle fencing and telecommunications. Many of the systems have been of an experimental nature, for developing and demonstrating the technology, but increasingly photovoltaic systems are being installed for sound commercial reasons, as being the most cost effective solution for particular applications.

The technical, economic, social and institutional factors associated with each application need to be carefully considered before any general conclusion can be made regarding the viability of photovoltaics. Obtaining reliable data on the performance of PV systems is not easy, as there have been only a limited number of published reports on actual field experience. It is also not easy to evaluate the economic prospects, as many systems have not been operating long enough to yield sufficient data on component lifetimes and replacement costs. Social and institutional factors have a major influence on the success or failure of projects to introduce new technologies into

the rural sector and many valuable lessons have been learned
in the course of implementing PV projects in different
countries.

### 3.7.2 The Technology

In general, the photovoltaic modules and arrays have
performed reliably with very few reports of failures or
significant degradation.   The experience to date has been
solely with crystalline silicon solar cells, both mono-
crystalline and semi-crystalline.   This is because most
manufacturers make module which meet the qualification and
performance requirements laid down by the Jet Propulsion
Laboratory (USA) or the CEC's Joint Research Centre (Italy).
The performance of the newer types of thin film amorphous
silicon solar cells, which are beginning to become available
for power applications, has yet to be evaluated.

Many early PV systems (i.e prior to about 1982) suffered
problems with power conditioning and control systems, such
as voltage regulators and maximum power point trackers.
This was due to a number of factors, including inadequate
weather protection, over complex design and not sufficiently
robust, both physically and electrically.   There is growing
evidence that the power conditioning and control equipment
now being supplied for use with PV systems is much more
reliable, but further development to improve the efficiency
of some devices, particularly inverters, is required.

For systems that incorporate batteries, it is vital that the
correct type of battery is chosen and that the sizing
calculation takes full account of operating conditions.
Automotive batteries are, in general, not suitable, as they
have limited life when subjected to many charge/recharge
cycles. A number of battery manufacturers have developed
special batteries for PV applications which offer long life
and   low   internal   losses   and   require   little   or   no
maintenance.

The   end   use   devices,   such   as   motor/pump   units   and
refrigerators, have generally had to be specially developed
for PV applications.   Many problems developed with the
earlier systems, often due not so much to faults in the
overall   concept   but   to   mistakes   in   the   matching   of
components, faulty operation and inadequate quality control.
New   products   with   better   performance,   reliability   and
durability have become available in recent years for all
applications of main interest in developing countries.

There has been extensive work to test and evaluate PV
pumping systems and PV vaccine refrigerators.   There is a
continuing need to update the results of previous projects
and to keep potential customers informed of the state-of-
the-art.   Similar testing and evaluation programmes are

needed for other PV systems, such as lighting systems, electrification systems suitable for households and institutional buildings (health centres, police posts, schools etc.) and systems for agricultural product processing such as grain mills.

### 3.7.3 The Economics

It is not possible to generalise about the economic viability of PV systems. Each application has to be considered on its merits, taking into account local conditions and the cost of alternatives. Although PV systems have high initial cost, they require no fuel and little maintenance and should last many years. In many remote areas, diesel generators, the main alternative to PV generators, would be impractical due to fuel supply costs and uncertainties, plus the problems associated with maintenance and the supply of spare parts. Even if diesel generators appear to be cheaper on a life-cycle cost comparison, it might be preferable to go for a PV system because of the operational advantages.

For some applications, PV systems are widely found to be competitive with the alternatives. For example, PV refrigerators for vaccines offer a lower cost per dose than kerosene refrigerators and allow a more effective immunisation programme to be mounted. PV systems for domestic lighting are also competitive with kerosene lamps and candles and, moreover, give a much better light.

PV water pumps may be cost competitive with diesel pumps for applications where the flow is low (less than 50 $m^3$ per day) and the head is low to medium (less than 20 m). The analysis is sensitive to the cost of diesel fuel and the life expectancy of the diesel pump. PV pumps are more likely to be viable for village water supply or livestock water supply applications, where social benefits are high, than for irrigation applications, where the extra value of the crop made possible by the irrigation water has to exceed the cost of the pump.

PV systems for telecommunications and cathodic protection are often found to be the cheapest option when small amounts of power (up to about 10 kWh/day) are required at sites remote from public electricity supplies. Depending on local circumstances, PV systems may also be the cheapest alternative for hazard warning lights, remote metering stations, navigation buoys and lighthouses.

### 3.7.4 Social and Institutional Factors

Experience has shown that PV systems are generally widely welcomed by the users, provided that they have been involved at an early stage in the planning process and have been given basic instruction in how to operate the system and carry out routine maintenance. Problems arise when a system is set down in the field by a research organisation and the local people are expected to use it without any real appreciation of what is involved and without proper arrangements for follow-up.

Unlike diesel systems or grid-supplied electricity, PV systems are very site and load dependent. Therefore, considerable experience and technical expertise are needed at the design and procurement stage in order to ensure the system will fulfil its expectations. A central organisation within the country concerned needs to be established with the necessary skills in-house to undertake the necessary design tasks and write system specifications in preparation for tendering procedures. The same central organisation can arrange for user training and technical support in the operation phase. The supply of spare parts, particularly where these have to be imported, is an important function of this central organisation.

There are four important institutional factors that contribute to the success of a PV application:

a)    Genuine involvement by the users from the planning stage onwards, with appropriate arrangement for a co-operative or other method of administering the local aspects (operation, maintenance, fund raising, user charges, etc.);

b)    Technical expertise at the system design and procurement stage to ensure that the system is compatible with local needs, solar resources and other technical constraints;

c)    Finance to meet the majority of the initial costs, after allowing for any government subsidies considered appropriate, with repayments spread over at least five years at a preferential interest rate;

d)    Arrangement for user training and technical support after installation to deal with faults that cannot be fixed locally and the supply of spare parts.

Although most developing countries have already established a department to undertake the necessary supporting activities as outlined above, many need technical assistance and finance from outside to help them identify and implement appropriate photovoltaic projects.

3.8 Conclusions

In recent years, several thousands of PV systems have been installed in developing countries to supply power for water pumping, refrigeration, lighting, communications and other applications. Many of these systems were part of development and demonstration programmes and, therefore, some operational problems were encountered. However, their installation has allowed valuable field experience to be obtained and present PV systems have been improved as a result of assessment of these development programmes.

At present, PV systems can be shown to be cost effective for vaccine refrigeration and telecommunications for off-grid locations and for water pumping, for human and livestock consumption, in many cases. The reliability and low maintenance aspects of PV systems makes them particularly useful in medical applications. As the cost of PV systems continues to fall, further applications will become cost effective.

PV systems are of high capital cost, compared to alternative power supplies such as diesel, although life cycle costs are often lower. Therefore, one important aspect of the introduction of PV systems is the provision of finance for their purchase, with low interest rates. This is becoming more easily available as financial and government institutions begin to appreciate the advantages of this fuel-free, pollution-free electricity source, especially for rural areas.

Provided that the users have been involved from an early stage, the PV systems are usually very popular. From a technical viewpoint, they are now seen as a standard engineering solution to many power supply problems in remote areas. Thus, the use of PV systems is a desirable route to the expansion of energy supplies in developing countries, leading to their social and economic development without the financial or environmental problems of imported fossil fuels.

## 3.9 References

3.1  "Small-scale solar-powered pumping systems: the technology, its economics and advancement". Main report prepared by Sir William Halcrow and Partners in association with IT Power Ltd., for UNDP Project GLO/80/003 executed by the World Bank, June 1983.

3.2  IT Power, "Solar Powered Pumping Systems: Their Performance and Economics". Report to World Bank (1986).

3.3  Mali Aqua Viva report No. 7, Activities from October 1979 to October 1981.

3.4   T.R. Miles Jr., "Economic Evaluation of Renewable
      Energy Technologies at LESO" for US-AID and
      Energy/Development International, December 1985.

3.5   "Small-scale solar-powered irrigation pumping systems -
      Phase 1 Project Report" by Sir William Halcrow and
      Partners in association with Intermediate Technology
      Development Group Ltd., for UNDP/World Bank, July 1981.

3.6   "Small-scale solar-powered irrigation pumping systems -
      Technical and Economic Review" by Sir William Halcrow
      and Partners in association with Intermediate
      Technology Group Ltd., for UNDP/World Bank, September
      1981.

3.7   J.P. Kenna and W.B. Gillett, "Solar Water Pumping - A
      Handbook", I.T. Publications, London, 1985.

3.8   P. Amado and D. Blamont, "Implementation of a solar
      pump in a remote village in India; economical and
      socio-economic consequences - three years of working
      experience", Proc. of Third International Conference on
      Energy for Rural and Island Communities on Energy or
      Rural and Island Communities, Inverness, Scotland,
      September, 1983.

3.9   R. McGown and A. Burrill, "Current Developments in
      Photovoltaic Irrigation in the Developing World",
      A.R.D. Inc., 1985.

3.10  D.R. Darley, "PV vs Diesel: A Grounded Economic Study
      of Water Pumping Options from Botswana", Massachussetts
      Institute of Technology, USA, 1984.

3.11  I.T. Power Inc., Proceedings of the Photovoltaics
      Information Symposium and Workshops, Nairobi and Chaing
      Mai, April 1986.  (Sponsored by the UNDP Energy
      Office).

3.12  IT Power Ltd., "An Evaluation of PV Applications for
      Developing Countries", UNESCO Report Ref. No. 85192,
      April 1986.

3.13  "Solar refrigerators for Vaccine Storage and Ice
      Making", World Health Organisation, EPI/CCIS/81.5,
      1981.

3.14  "Purchasing Guidelines and Product Information Sheets",
      World Health Organisation, EPI/CCIS/85.4, 1985.

3.15  "The Cold Chain Product Information Sheets", World
      Health Organisation, SUPDIR 55 AIT 5, 1985.

3.16 A. Derrick and J.M. Durand, "Photovoltaic Refrigerators for Rural Health Care - Experience and Conclusions". Proc. of the UK-ISES Conference Solar Energy for Developing Countries - Power for Villages, London, May 1986.

3.17 A.F. Ratajczak,"Photovoltaic-Powered Vaccine Refrigerator/Freezer Systems; Field Test Results, NASA-LeRc, 1985.

3.18 B. McNelis and J.M. Durand, "Photovoltaic Refrigerators and Lighting Systems for Zaire", Proc. of 6th EC Photovoltaic Solar Energy Conference, London, April 1985.

3.19 "Pre-investment Report on Solar Photovoltaic Applications in the Health and Telecommunications Sectors, The Gambia", UNDP/World Bank, March 1985.

3.20 G.H. Kinnell, "Solar Photovoltaic Systems in the Development of Papua New Guinea", Proc. of the 4th EC Photovoltaic Solar Energy Conference, Stresa, May 1982.

3.21 K. Maleva, "Feasibility Assessment for Photovoltaic Cells Replacing Kerosene Lighting in Papuan Villages", Report No. 7/81, Energy Planning Unit, Dept. of Minerals and Energy, Konedobu, Papua New Guinea.

3.22 H.A. Wade, "The Use of Photovoltaic Systems for Rural Lighting - An Economic Analysis of the Alternatives", Proc. of the Solar World Congress, Perth, August 1983.

3.23 "Renewable Energy in French Polynesia", by SOL E.R. and CEA-GER, Pappeete, French Polynesia, 1985.

3.24 Zhu Gangi, "Recent Experience on Solar Energy Utilisation in China", Proc. of UK-ISES Conference C42, Energy For Development - Where are the Solutions?, Reading, UK, December 1985.

3.25 M.R. Starr, "Photovoltaic Prospects for Rural Electrification", Proc. of INTERSOL 85 World Solar Energy Conference, Montreal, 1985.

3.26 M.R. Starr, "Rural Electrification - Solar Versus Grid Extensions - Updating the Economics", Proc. of UK-ISES Conference, Solar Energy for Developing Countries - Power for Villages, Reading, May 1986.

3.27 "Solar Village Indonesia", brochure published for TUV Rheinland Institute for Energy Technology, W. Germany, and BPPT Badan Pengkajian Dan Penerapan Teknologi, Indonesia.

3.28 S. McCarthy, G.T. Wrixon and A. Kovach, "Data
     Monitoring of the Photovoltaic Project", Presented at
     the First Working Sesion of the European Working Group
     on Photovoltaic Plant Monitoring, Ispra, Italy,
     November 1985.

3.29 "Water Disinfection System and Cold Store", Design
     Report Presented by Pragma at Contractors Meeting of
     the Commission of the European Communities Photovoltaic
     Pilot Projects, Nimes, France, April 1983.

3.30 B. McNelis and S. J. Lancashire, "Solar Powered Water
     Treatment/Delivery Systems for Indonesia - A Cost
     Benefit Analysis", Proc. Aquarius '88, Jakarta, 1988.

3.31 D.A. Hughes and A.B. Wood, "Jamuna River 230 kV
     Crossing - Bangladesh: II. Design of Transmission
     Line", Proc. Instn. Civil Engineers, Part 1, Vol. 76,
     pp 951-964, November 1984.

# Photovoltaic Systems for Professional Applications

Alan Dichler

Solapak Ltd., Factory Three,
Cock Lane, High Wycombe, Bucks.

## 4.1 Introduction

No longer is a solar generator a 'new' technology. It may come as a surprise that the photovoltaic effect was observed around 150 years ago and devices using this effect were first manufactured for general usage some thirty years ago.

However, since the early days of solar there have been many advances in the active device itself in terms of reliability, quality and efficiency, not to mention a fundamentally important parameter - cost. The application in which solar is most commonly known is in space exploration but, for the majority of us, the interest in solar systems lies in terrestrial applications and particularly in those where there is no conventional electrical grid connection. Of these, the most common application is in the telecommunications field where the requirement for remote, reliable, fuel-less power systems that a solar generator offers are most suited. However, other applications are springing up daily and cathodic protection, railway signalling, data collection, telemetry, vaccine storage refrigeration, ventilation and water pumping are but a few of today's applications.

This chapter first discusses aspects of photovoltaic systems, such as battery storage, system controllers and system design criteria. This is followed by a review of the major applications of photovoltaics which are presently cost effective, including telecommunications, cathodic protection and rural power supplies, and the final section deals with photovoltaic hybrid systems, i.e. those in which photovoltaics is coupled with one or more other electricity sources, to give greater flexibility or reliability of supply. A description of the solar-wind hybrid system installed by Solapak at Milton Keynes is included as an example of a practical system.

## 4.2 Photovoltaic Technology - Minimising Generator Costs

### 4.2.1 The Photovoltaic System

A photovoltaic generator system consists of three major sub-systems - the solar array, the control unit and the storage battery system (Figure 4.1). The efficiency and cost effectiveness of the PV generator is obviously dependent on

the quality and specification of each of these sub-systems.
About ten years ago, photovoltaic cell processing techniques
increased cell efficiency to levels acceptable for
professional systems. Telecommunications is the largest
market area for photovoltaics and, very rapidly, the
advantages of reliable, clean, virtually maintenance free,
silent power sources were accepted by the telecommunications
equipment manufacturers and photovoltaic systems became
recognised and accepted as a viable source of electric
power.

Technical developments in control equipment implementation
have aided this acceptance of photovoltaics. The advent of
solid state circuitry offers the same advantages to
photovoltaic system controllers as it has to
telecommunications equipment designers. There has been an
increase in unit reliability and efficiency and system
controls are now established as a very important part of the
solar generator. Selecting the right control unit for each
photovoltaic system is important.

Developments in control circuitry have meant that fault
detection, better monitoring and digital metering of the
system has become easily available, and, as authorities have
used these circuits to check the actual system performance
against prediction, their confidence in solar system
reliability has increased. In the last two years, new
developments in maximum power point tracking and the
treatment of system operational characteristics have enabled
power gains of 20-25% to be realised over traditional
methods of control. For example, the Solamax controller
(Ref. 4.1, Figure 4.2) actually reduces solar array costs by
following the peak power point and conditioning the array to
operate close to this, hence improving the array
performance. The improved array efficiency obtained with
Solamax is directly reflected in the quantity of solar
modules required and, thus, the size of the support
structure, which in turn is reflected in reduced
installation and civil work costs.

### 4.2.2 Battery Storage

Energy storage is a necessary part of any stand alone solar
generator. In the field of rechargeable batteries, two
electrochemical systems are of technical importance, the
lead acid battery with sulphuric acid as electrolyte and the
nickel cadmium battery with potassium hydroxide as the
electrolyte. The requirement for performance versus cost
supports the selection of a lead acid battery for
photovoltaic systems. Lead calcium batteries have been used
extensively, because their low self discharge rate made for
efficient solar system design. However, such batteries have
limited cycle lives and low antimony, tubular plate
batteries, which have a very good self discharge and cycling

Figure 4.1    This photograph show a cathodic protection
              system installed in the Middle East. It
              illustrates the three major subsystems of a
              photovoltaic system - the PV array, the
              control units and the battery system.

Figure 4.2    This figure illustrates a PV system control
              unit (in this case, the Solamax controller),
              which provides maximum power point tracking
              and a monitoring facility.

properties, are now often chosen. The use of these low antimony batteries permits the system designer to minimise the battery capacity while still maintaining a high level of security. Photovoltaic charging regimes are now being considered in battery design and battery manufacturers are becoming concerned with construction of batteries suited to solar applications.

Features looked for in a photovoltaic battery are:
- long service intervals (up to 3 years)
- low self discharge (less than 3% per month)
- high cycle life (about 1000 cycles per 80% depth of discharge)
- resistance to overcharge
- high charging efficiency.

Additionally, the availability of more detailed battery data has meant that the design of the solar generator can take into account the needs of the storage battery and, with the assistance of suitable control units, the system can be selected for the most efficient battery treatment. This will extend battery life and reduce site maintenance visits, therefore minimising the attendant maintenance cost, which is particularly important for solar systems at very remote sites where the cost in man hours and transport can be very high.

With regard to other battery types, the latest contender for solar generator systems is the sealed maintenance free battery system. In theory, this is an ideal component in that it offers a maintenance free or very minimal maintenance capability which matches that of the other parts of the solar system. It has the great advantage that it is a factory sealed unit so the installer is not beset by problems of supplying acid, filling, commissioning etc. However, this battery has only been around in solar applications for the last few years and field experience is limited. Certainly the controller used with this type of battery needs to be very responsive to changes in charge voltage, so as to limit any overcharge condition. A disadvantage of this battery type is its limited ability to withstand abuse during fault conditions and the difficulty in readily determining state of charge. They also have a higher first fit cost.

### 4.2.3 Solar Array and Structure

Solar cell and module technologies have also developed. Techniques for cell cutting have allowed an improved module packing density. Round monocrystalline cells can be economically cut to a square cell with a correspondingly improved output per unit area (Figure 4.3). In the early 1980's, the polycrystalline cell offered cell efficiency of 9.5% and the manufacturing techniques meant that no additional cell cutting was required. The last two years have seen the advent of high efficiency monocrystalline cells, with increased cell efficiency of 12%, and, when combined with the improved cutting techniques, a packing density of 115 W/m$^2$ is now available. The latest developments are in thin film technology which has reached a stage where larger, higher power modules are viable and 1985 witnessed the introduction of the first commercially available amorphous silicon solar modules, although they are still not usually suitable for professional systems due to their low conversion efficiency.

Structural design improvements have aided the effectiveness of solar generator systems. In the early 1980's, reliable tracking devices improved solar system performance by maximising the solar insolation falling on the surface of the solar module. Static structure design has improved as new trapezoidal and tubular type structures increase the speed of installation and reduce civil costs.

In areas of high vegetation growth, shadowing of the solar array is possible and causes a readily observed reduction in output. To minimise this possibility, professional solar systems companies will generally recommend that the solar array be lifted, say 1.2 metres above ground level and, in some cases, even as much as 2-3 metres. Unfortunately, raising the array can have other attendant problems with regard to wind loading effects and proper structural design is essential if the system is to survive high wind speeds often found at remote sites.

### 4.2.4 Control Units

With the ongoing demand for more information regarding the performance of a solar generator system, the control unit can now offer a wide range of different characteristics and facilities that were, even a few years ago, unheard of. For example, it is not uncommon these days for the control unit to be required to perform power tracking of the solar array, to include temperature compensation, to provide battery monitoring with high and low voltage alarms and even, in some cases, to include digital metering and monitoring to record the solar insolation present on the site and the charge which flows to and from the battery system. This is in severe contrast to a very early control system, which in its most basic form was just a diode to prevent the solar array discharging through the battery system in periods of darkness.

It should be remembered, however, that the control unit on a solar system is there purely to maintain the battery in the highest state of charge possible without overcharging, or, in the other extreme, leaving the battery in a very low state of charge. Additional monitoring/metering facilities

serve to provide the user with information with regard to
the overall system performance at any one period of time.
Usually this control equipment comprises a solid state
mechanism rather than the more unreliable relay type
regulators which have also been used. Most control units
nowadays offer on-site circuit card replacement and the best
control units have simple circuitry to enable circuit board
repair to take place in local workshops.

### 4.2.5 System Design

On the face of it, solar generator design seems a fairly
straightforward process largely aimed at estimating the
average output from the solar system based on a given
radiation pattern for any one particular site area. However,
this calculation is not a simple one and involves a
considerable amount of experience in deciding which factors
are the most significant within the overall design concept.
For example, these may include:-

-       the nature of the load, be it constant power, constant
        current or constant resistance
-       the acceptable safety design factor to meet a
        particular load security requirement
-       treatment of the basic insolation data used in the
        solar system design
-       the respective percentage levels of diffuse and direct
        sunlight
-       tilt angle of the solar array
-       temperature derating for the solar array
-       peak power tracking of the solar array
-       percentage of the load consumed during the day and
        night
-       battery efficiency prediction
-       cost optimisation of the solar array versus the
        required battery capacity and balance of system
        components.

Detailed consideration of these parameters goes beyond the
scope of this chapter, but it should be stated that a
successful solar generator system, like any system, is not
just a collection of components, but a matched design
system, often including the load equipment, with attention
being paid to even the smallest detail to ensure a reliable
and efficient result. In particular, the experience and
history of those carrying out the solar generator system
design is an important factor.

### 4.2.6 Cost Effectiveness

On the basis of initial capital cost, solar generators are
often still rejected in favour of conventional generation by
diesel generator sets. In many instances, if the real
operating cost over the lifetime of the equipment involved

was considered, then this decision would be reversed. Even with a simple analysis, it can be found that solar energy is extremely competitive when the total operating cost per annum is considered. This is especially true for remote sites.

Although there is no one rule governing when a solar system is effective, it can generally be stated that the more remote the site and the better the solar radiation, then the more likely it is that solar will be cost effective in comparison to other forms of generation. This is especially true in the communications field. In remote sites, solar for a microwave link can certainly be cost effective when compared to diesel generator sets, with a payback period of 3-5 years.

As technology advances even further, we can expect a further decrease in the cost of the solar, but the cost of control equipment and batteries will inevitably rise in the future. For the next few years, however, it is anticipated that the real cost of solar generator systems will continue to fall, with the possibility of amorphous technology becoming accepted for use in the professional systems arena. The following sections outline typical application areas where solar electric systems are already being used to advantage.

### 4.3 Cathodic Protection

Photovoltaic systems have found wide acceptance for cathodic protection of pipelines and other metal structures (Figure 4.4). The PV system is particularly well suited to this application for two main reasons. Firstly, power is required in remote locations through which pipelines are installed where it can prove difficult to install mains electricity or diesel generators. Secondly, the technique of cathodic protection, involving the maintenance of the pipe to be protected at a negative potential with respect to the surrounding earth or atmosphere, requires a DC source of electricity. This means that a PV system can be used directly, although it is common to include a DC-DC converter to allow the necessary current/voltage conditions to be produced.

To provide system flexibility and standardisation, control equipment can be made such that the required output conditions (voltage and current) can be "dialled in", and the equipment then left to maintain these conditions automatically. Virtually any combination of current and voltage can be selected within the overall available power. Such equipment has been a standard part of the Solapak cathodic protection equipment range for many years and employs DC-DC switch mode conversion principles.

Figure 4.3          Presently available modules use square cells
to give a very high packing factor, as
illustrated in this photograph, and, hence, a
high power per unit area of module.

Figure 4.4          Cathodic protection system installed on a
pipeline in North Africa.

Configuring the array at a standard of, say, 24V offsets some of the problems of wiring compared to lower voltage (say 6V) high current supplies and solar system design can be optimised at this voltage level. Voltage down-conversion is undertaken by the output control to produce the lower voltage, higher current required by the load. Such conversion can be "open loop" where output is set for one specific load condition or "closed loop" where ground bed potential is monitored by a half cell and the current adjusted accordingly. Such a control system can be expanded by providing new converter elements as the need arises.

DC-DC conversion equipment is now widely used for equipment power supplies and has proved to be reliable, efficient and cost effective without the need for high power elements to dissipate excess power used in conventional supplies, so that equipment is far less cumbersome and arranged in plug-in modules, easy to fault and maintain.

## 4.4 Telecommunications

Telephone lines and transmitters used in communications systems are so costly that most telecommunications equipment development is concerned with ways of transmitting the maximum amount of information through the smallest possible number of channels and with the smallest possible size of transmitter. The system designer needs to establish the best compromise between cost and method of transmission. There have been a number of significant steps which have moved towards this optimisation of power and increased equipment efficiency.

Communication by satellite and microwave radio circuits is now widespread and the latest equipment technologies are much less onerous in terms of the size of power supply they require. The pioneering work undertaken by several major radio equipment companies, even with power hungry equipment, has demonstrated the viability of solar and very high reliability figures have been demonstrated. It is to be expected that others will follow and the telecommunications industry in all its forms both is now, and forecast to be, the major market sector for solar applications. An illustration of a typical PV powered telecommunications system is shown in Figure 4.5, whilst Figure 4.6 shows telecommunications compatible control equipment.

New technologies such as optical fibres are increasingly competitive as the higher bit rate capacity allows them to reach a very long span without resorting to a regenerative repeater. Digital technology has also made significant advances into the field of telecommunications since digital switching transmission clusters offer reduced equipment and elimination of multiplex, and therefore a reduction in overall power requirements.

Figure 4.5        A Solapak solar generator powering a
                  microwave repeater link in West Africa.

Figure 4.6        Telecommunications compatible control
                  equipment for a PV system.

However, perhaps the most significant advance in communications equipment technology in the last two decades is the advent of the solid state design technology which has meant an overall improvement in conversion efficiency. Twenty years ago, most communications equipment was valve powered and run from inefficient high voltage AC power supplies about 30% power efficient. The new solid state technology permits 24V DC and 48V DC systems which are approximately 50% power efficient. This development is a particular interest within the field of photovoltaics as the inverter inefficiencies of producing an AC supply from a DC power source are also removed, giving an additional 20-30% power saving. Antenna development has also given a boost to communications technology as aerial design and impedance matching has reduced the antenna losses and less power is now required to propagate a given path.

Each of these advances in telecommunications technology increases the effectivness of the power path and reduces the number of repeater sites, increasing the system efficiency and thus reducing the size of power system required. Communication and photovoltaic equipment as a total system rather than two (or more) individual elements enables a systems integration hitherto not possible, and the effectiveness of this approach is now becoming recognised such that this applications area is, in fact, increasing as the economic viability of installing solar powered communications systems increases.

### 4.5 A Key to Rural Development

Rural development has now become an important factor in overall Government planning and most developing countries have begun to implement major programmes in the field to improve productivity and living standards. An essential for any balanced development is a reliable electricity source. In urban areas this is achieved by mains generated power using coal, oil or diesel fuels. However, the cost of extending the urban electricity grid into rural areas is often prohibitive due to the high capital investment involved. To date, diesel generation in rural areas has been the norm; however, diesel sets have an inherent disadvantage in that they must be fuelled, serviced and maintained regularly, causing logistic and manpower problems which may not be easily overcome. Additionally, the ever increasing energy bill has an effect on the financial resources of such countries. Recently however, studies have shown the feasibility of utilising solar energy to provide a rural electrification network for communities. Chapters 2 and 3 in this book have discussed the relative costs of PV and other power sources in developing countries and detail the major uses of PV systems in these countries.

The application areas where solar can aid the expansion of developing economies can be summarised as:-

*       Health care and hygiene

*       Education and social services

*       Agriculture and water resources

*       Communications (telephone) networks

Because of the modular nature of solar photovoltaic systems, the expansion of this applications area can be logically planned, allowing programmed increases in trade, whilst at the same time reducing the financial burden of importing fossil fuels.

In implementing a programme of this nature, the objective must essentially be to provide the rural community with basic items which will enhance their quality of life. This can be achieved easily and cost effectively and a number of examples follow. It must be noted, however, that unless a basic infrastructure is in place for operating and servicing these 'improvements', their long term impact is likely to be in doubt.

### 4.5.1 Refrigeration

In remote locations many unnecessary illnesses and deaths occur due to the lack of even the most rudimentary medical facilities, so villagers are enforced to walk or be carried to the nearest medical centre, which may be miles away. Until now, the primary health programmes have been seriously hampered by the difficulty of storing vaccines, serums and other medical supplies onsite.

As discussed in more detail in Chapter 3, solar energy has now become practical and a cost effective means of powering refrigerator/freezer systems for use in dispensaries, clinics, hospitals and homes. Complete fixed or portable systems, which offer maintenance free refrigeration and 24 hour continuous operation, are available. These systems provide a simple yet significant contribution to health care which can now be made available to any community.

### 4.5.2 Water Treatment

Health can also be seriously affected by drinking contaminated water, which, at best, can cause serious disruption within the community. It is now possible, by using a water treatment unit, to provide potable water. These units can purify by either chemical dosing or UV treatment of water delivered from either pumped or gravity fed sources. They are usually transportable, rugged and totally automatic - ideal for village use, in fact.

### 4.5.3 Educational Systems

These systems have been designed specifically to help improve literacy standards and to provide the most effective means of communicating information on a range of basic subjects through the use of television and radio. Teaching by means of television or radio allows a fast and effective way to reach a large and dispersed audience and to provide educational, social and recreational programmes at reasonable cost - since most countries already have an existing broadcast network. A wide range of solar powered television units are available, together with DC power packs for radios and cassette recorders.

### 4.5.4 Water Pumping

Though the market for solar water pumping systems has not grown in accordance with the predictions of the early 1980's, solar energy is now providing a reliable water supply for domestic consumption and agricultural use in rural communities. The systems are suitable for both shallow and deep wells where flows of up to 25 $m^3$ per day can be easily achieved at depths of up to 120 m or more. The detailed case for the use of PV water pumping was presented in Chapter 3.

### 4.5.5 Agricultural Sprinklers

Cash crops require a steady, controlled flow of water from above ground rather than ditch irrigation, which is more commonly used for the main crop. To meet this demand, a solar powered water sprinkling system can very effectively be designed. Multiple motor/pump systems feed water through small-bore distribution pipes to the sprinklers. The flow and droplet size is adjusted on the sprinkler heads to provide the correct spray pattern.

The use of sprinklers regulate the water flow, thus allowing the plant to absorb only the quantity it requires, leaving the excess water to evaporate into the air. This increases the crop yield and makes better use of the land area since water logging and seed decay is reduced to a minimum.

### 4.5.6 Portable Crop Sprayers

Conventional crop sprayers are expensive, cumbersome and time consuming and, as a result, many rural farmers are reluctant to use them. A novel, light and portable solar powered insecticide and fungicide sprayer has been designed. This can be used with ease and allow the motor and the unit

to take all the hard work out of spraying and field trials
are yielding very good results.

### 4.5.7 Electric Fencing

Overgrazing and damage to crops caused by livestock present
serious problems which need not arise if the fields and
pastures are properly enclosed. However, even conventional
fencing enclosures may be destroyed or trampled by
livestock. Users have long since recognised the advantages
of the electrically charged fence which can separate animals
from cultivated areas without undue damage to either.

Solar energy is now powering electric fences for livestock
separation, protection of stored feed, pasture rotation etc.
The unit eliminates the need for expensive battery
replacement or inconvenient battery charging. In fact,
recovery costs in terms of savings on batteries and time can
be realised within two years. The sun's energy powers the
fences by day while a storage battery provides power at
night, thus ensuring a continuous electrical pulse through
the fence and minimal maintenance. The fence drive is fully
weatherproof, portable and suitable for installation
anywhere providing power  for up to 40 km of fence in dry
regions.

### 4.5.8 Water Aerators

As many rural and coastal communities are dependent on fish
farming as a viable source of income, solar can provide the
necessary energy to run the tank aerator motors reliably and
economically to ensure a steady supply of healthy grown
fish.

### 4.5.9 Fan Ventilation

Ventilation systems can be essential, not only for human
comfort, but more importantly for reasons of economy.
Without adequate ventilation, in agriculture for example,
valuable livestock and grain may suffer and deteriorate
needlessly. Equally, in industry, the health of the
workforce can be seriously affected by poor ventilating
factors. There are four main areas where solar powered fan
ventilators can be effectively and economically put to use:

*    Vehicles - both land and marine
*    Livestock and grain storage
*    Sanitary outbuildings
*    Domestic buildings/extensions

### 4.5.10 Domestic and Street Lighting

The installation of solar lighting systems can bring this
convenience to even the remotest location. High efficiency

fluorescent lights are usually used because their output is equivalent to traditional globe lights of five times the power rating. Street lighting units are also available, being sealed against moisture, and can be mounted either on the outer wall of a dwelling or on poles along a street, on roundabouts, obstacles etc.

### 4.5.11 AC Power Packs

In some areas of the world, the availability of mains generated power cannot be entirely relied upon. To keep essential services (vaccine storage banks, life support equipment, emergency lighting, signalling systems, water pumping and treatment units and many others) operating until the electricity is restored, a standby power source is essential. For situations such as these, solar powered AC power packs are available, which can be directly coupled to the AC load equipment. These power packs include an inverter to derive the AC power from the battery storage.

## 4.6 Hybrid Systems

### 4.6.1 Introduction

Within the PV industry, the term hybrid is most commonly applied to a solar diesel cogeneration system where the introduction of a diesel engine is readily understood, provides good emotional security and yields a system having a relatively good cost index when compared to a stand alone solar installation. Hybrid systems using wind turbine generators are also known, but historically at least, attract a lower cost/reliability ratio than solar/diesel plant.

Two further power sources, the thermoelectric generator and the closed cycle vapour turbine have been considered for use but have also been labelled "unreliable" or criticised for having poor fuel consumption. But why hybrid at all? A brief review is offered in this section, which then aims to examine the role of the PV element when used in conjunction with other "conventional" power sources.

### 4.6.2 General Assessment

When dealing with power systems, there are four criteria that will certainly apply:

1)   The probability that the load power requirements will be met;
2)   Reliability;
3)   The installed cost;
4)   The recurring cost.

Whether power requirements will be met depends upon fuel availability, assuming a properly designed and engineered system.    If fuel is freely available to the generating plant, then system designers can assume no additional intermediate storage and the plant can be rated for the maximum load demand alone.  System surpluses can be reduced or eliminated.   This would be the case with conventional power sources.   However, when the availability of input energy is variable, intermediate "safety" storage has to be provided and the rating of the generator, size of the storage and surpluses will reflect that variability. Knowledge of the energy input and plant operating history are factors used when assessing reliability.  The installed cost usually influences the purchase decision without necessarily regarding recurring costs, since these are often met from a separate budget allocation.

Power supplied from the mains grid is usually a preferred choice since, providing the supply feeder already exists, this power is relatively inexpensive.  However, there are many areas where continuity of the existing mains supply cannot be guaranteed, and there are significantly large areas of the world without a mains supply at all.  If a reliable power source is required in either of these areas, then diesel engines have been a usual first choice.  These usually prove to be satisfactory provided that the maintenance routines are adhered to and providing that the infrastructure exists to ensure adequate supplies of the correct fuel.  Diesel generating sets are relatively cheap to buy and so have found wide application with many installations featuring two, often three, sets if no break in supply is to be ensured.  For power requirements above 10 kW, mains or diesel equipment is still the preferred economic choice.    However, at lower power levels, alternative sources are becoming more widely used, namely:

- wind turbine generators
- solar photovoltaic arrays
- thermoelectric generators (TEG)
- closed circuit vapour turbines (CCVT).

Wind turbine generators rank top of this list simply because they are a source of AC power as is the usual output from diesel generating sets, whereas the other three are usually DC power sources.  Wind turbines and solar photovoltaic arrays are "fuel free" whilst the TEG and CCVT require a supply of good quality fuel.  Careful consideration of unit capital cost is necessary for TEG and CCVT equipment, since these units are available only in finite power ratings. A load of 750 W may require a 1 kW rated generator (illustrative figures only) for instance.

A comparison of some hybrid systems is given in Table 4.1, which considers their suitability for installation in developing countries.

In broad terms, wind turbines and solar photovoltaic arrays offer the user freedom to choose a location best suited either to the application or to the wind/solar availability. These systems are constrained by the variable nature of the input energy. Thus a sufficiently large storage system needs to be included if continuity of output is to be ensured. Both the TEG and CCVT deliver known and constant output energy, so the need for storage is minimised or even eliminated. Both systems are constrained in regard to their location which must be accessible for fueling visits.

Photovoltaic, TEG and CCVT systems have exhibited very high reliability in use (Ref. 4.2), and an increasingly attractive proposition is to take into consideration a mix of power sources to create a co-generation "hybrid" system, with the worst attributes of each power source being offset by the best attributes of the other. Solar photovoltaic (PV) systems have been demonstrated to be reliable and cost effective (Ref. 4.3), and stand alone PV systems (non hybrid) will meet load demands up to 1 kW fully. Also, the recurring costs are low, typically one tenth of those of a diesel generating set. However, stand alone solar system design demands a significant "oversize" on safety factor. Adding a diesel set can extend the load capability by some 50% and, because of the small quantity of fuel involved, not seriously compromise the choice of location. Combining solar and wind takes advantage of sites where perhaps the winter wind and summer sun complement one another to increase the annual average power output of the system, and the use of TEG and CCVTs can bring benefits to a stand alone solar PV system by reducing the battery storage capacity required. Displacing battery storage by fuel storage can extend the battery operating life and so increase the interval to battery replacement.

Hybrid systems might conveniently be assessed in three categories such as outlined in the next three sections.

### 4.6.3 The Easy Systems

The so called "easy" systems relate to those where the input conditions for one of the two (two of the three etc.) power sources is non-variable and known. As the cost of photovoltaic plant reduces, it is expected that TEG, CCVT and perhaps some mains linked systems, will become supplanted totally by solar stand alone systems. However, there are applications where solar would complement an existing installation (power upgrade) or where diesel run time could be reduced by the addition of solar whilst still maintaining power availability. This could be particularly

Table 4.1        Comparison of various hybrid solutions

|                                              | PV + Diesel Genset | PV + TEG          | PV + CCVT | PV + WIND |
|----------------------------------------------|--------------------|-------------------|-----------|-----------|
| Fuel Supply                                  | Diesel             | Propane/ Butane   | LPG       | N/A       |
| Availability of Fuel                         | Good               | Poor              | Fair      | N/A       |
| Fuel Consumption                             | Good               | Poor              | Poor      | Nil       |
| Cost per 1000W Output                        | Low                | High              | High      | High      |
| Availability of Spare Parts                  | Good               | Fair              | Poor      | Poor      |
| Suitable for High Power Output 1000W+        | Yes                | No                | Yes       | Yes       |
| General Suitability for 3rd World Countries  | Good               | Poor              | Poor      | Fair      |
| Ease of System Design                        | Good               | Good              | Good      | Fair      |

relevent where insolation data is not well known or where the site is prone to extended periods of mist, cloud or ice cover etc.

### 4.6.3.1 Solar-Mains

In its simplest form, this configuration is used to back up load normally powered from the mains grid or utility. A photovoltaic plant will extend the fixed period of mains outage, otherwise obtained from a battery float charged from the mains supply whilst it is present. Here, the solar element is usually designed to generate sufficient energy to provide load continuity during mains outages and its modular nature allows expansion at any time. A usual extension to the basic principle, provided local regulations apply, is to use the power generated by the solar to feed back into the grid or other distribution network for the time it is not required by the load. Such installations are, however, unlikely to be in common usage until the new cell technologies are mastered technically and commercially exploited at costs of $1/Watt or less, perhaps by the mid 1990's.

### 4.6.3.2 Solar-TEG or CCVT

The largest single TEG has approximately 100 W output so that the larger loads would require several TEGs in parallel, whereas CCVT units are available up to 3 kW. The capital and installed costs of a TEG are similar to that of solar in the order of £100/continuous Watt, with CCVT plant attracting about half this, but fuel and maintenance adds further to this figure. In cathodic protection applications, it is often possible to fuel the TEG direct from the (gas) pipe line it is protecting. In other applications, the cost of purchase and distribution of fuel to the TEG/CCVT sites would increase the cost per Watt figure noticeably. Photovoltaics could be used to deal with the summer load, completely conserving fuel supply, perhaps by turning it off altogether. Automatic start facilities could be built into the solar control to restart the TEG/CCVT at the onset of autumn. Monitoring devices could signal a start failure early enough to allow corrective action before the system was fully depleted. With this regime, fuelling could reasonably be expected to be once per annum at most, which could be effected during the maintenance visit.

### 4.6.4 The Intermediate Systems

The title "intermediate" system is applied to plants having a known and non-variable maximum power capability but which also lends itself to having one power source switched on and off as required by the system control. This is characterised by the solar-diesel hybrid, but solar-TEG/CCVT combinations also come within this category.

4.6.4.1 Solar-Diesel

Now gaining in popularity, the solar-diesel hybrid has much to offer to a power plant user. Essentially, this hybrid is designed to meet the load demand by balancing the cost of the solar element against the number of starts and running time (and hence fuelling and maintenance costs) of a suitably sized diesel engine. Experience has proven that this is not as easy as might be considered at first thought. The solar element has to be used to its fullest capacity and the diesel must be started and run at regular intervals and must always be electrically loaded. All needs must be automatically accommodated by the hybrid control without any intervention by the user/operator. Figure 4.7 illustrates how a solar-diesel hybrid system may be employed to fulfil a substantial load demand. Note especially the absence of power surpluses.

Since installation costs can be considerable, the time on site should be minimised. It is convenient therefore to specify that the building which houses the diesel also mounts the solar array and accommodates the system battery. For remote locations, it would be ideal for a pre-assembled pre-tested building to be shipped, and containerised solar-diesel hybrid systems are now a practical reality for loads up to 2000 W continuous. Typical costs for such a system are about £80/W. A comparison of costs in relation to a stand alone solar PV system for a site with road access is given in Figure 4.8.

Figure 4.9 shows, in more detail, a typical design for a climate where the solar/diesel contributions are more evenly matched. This is a design analysis for a 1440 W continuous load. The solar array contributes 75% of the yearly power consumed by the load whilst the diesel contributes the balance in 66 starts, 12 of these primarily for excercise. The total running time of 795 hours per year is a very acceptable figure, giving good engine life and one fuelling visit. Surpluses can be seen to be very low indicating good utilisation of the equipment. Trials of this type of equipment have yielded very encouraging results.

The container concept could be extended to include the user's equipment. If there was, therefore, a need for temperature control, air conditioning could be applied to the container and the system designed to run this from array surpluses during the peak summer period. A typical design analysis is given in Figure 4.10. This shows how the air conditioning run time might vary with changes in external temperature. The resulting load on the power system can

**Basis of design**
Solar array designed to meet summer load
(with minimum wasted surplus)
Diesel meets shortfall in other seasons
Small battery bank to absorb hourly fluctuations
(to improve efficiency by minimising part-load operation)
Surplus power utilised elsewhere

**Primary design parameters**

| | |
|---|---|
| Solar array: | 5.5 kWp |
| Diesel capacity: | 55 kW maximum |
| Battery storage: | 28 kWh |
| Inverter capacity: | 25 kW minimum |

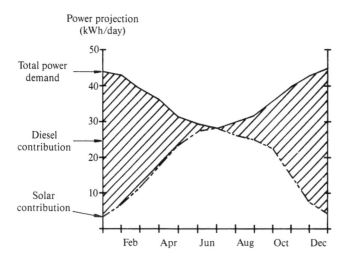

Predicted performance

| Demand (kWh/day) | 43 | 39 | 36 | 31 | 29 | 28 | 29 | 31 | 35 | 39 | 42 | 44 |
|---|---|---|---|---|---|---|---|---|---|---|---|---|
| Solar (kWh/day) | 7 | 11 | 18 | 23 | 27 | 28 | 25 | 24 | 22 | 14 | 7 | 3 |
| Diesel (kWh/day) | 36 | 28 | 18 | 8 | 2 | 0 | 3 | 7 | 13 | 25 | 35 | 41 |
| Surplus (kWh/day) | 0 | 0 | 0 | 0 | 0 | 1 | 0 | 0 | 0 | 0 | 0 | 0 |
| Sun (Hrs) | 1.7 | 2.5 | 3.7 | 4.7 | 5.3 | 5.5 | 5.1 | 4.9 | 4.4 | 3 | 1.7 | 1.1 |

Figure 4.7   Typical characteristics of a solar–diesel hybrid system (© 1987 Solapak Ltd)

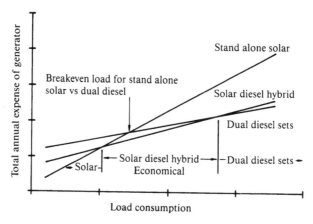

Figure 4.8   Comparison of costs for stand alone PV systems and PV diesel hybrid systems
(© 1987 Solapak Ltd).

**Indicative air-conditioning loads**

| | |
|---|---|
| Based on container dims: | 6 ft × 6 ft × 8.5 ft |
| with thermal coefficient: | 1.7 U |
| and construction: | shaded top |
| | |
| Air conditioner: | 240 Volts, 8 Amps total |
| Typical compressor operation: | 20 mins on 10 mins off |
| | |
| Internal dissipation (continuous): | 900 Watts |
| Internal temperature ceiling: | 30 °C |
| Max. temperatures: | 0° above published max |

| Mth | Daily temps | | Estd Hrs with external temp in range: | | | | | | Indic A/C Dly Opern | | |
|---|---|---|---|---|---|---|---|---|---|---|---|
| | Max | Min | 25–30 | 30–35 | 35–40 | 40–45 | 45–50 | 50–55 | Hrs | AC AHs | DC AHs |
| Jan | 28.3 | 18.9 | 9.0 | 0.0 | 0.0 | 0.0 | 0.0 | 0.0 | 6.9 | 45.9 | 574.3 |
| Feb | 28.9 | 19.4 | 11.0 | 0.0 | 0.0 | 0.0 | 0.0 | 0.0 | 7.3 | 49.7 | 621.4 |
| Mar | 33.9 | 22.2 | 8.0 | 9.0 | 0.0 | 0.0 | 0.0 | 0.0 | 10.4 | 70.0 | 875.5 |
| Apr | 37.8 | 25.6 | 9.0 | 8.0 | 7.0 | 0.0 | 0.0 | 0.0 | 13.3 | 89.3 | 1116.7 |
| May | 42.2 | 30.0 | 0.0 | 11.0 | 6.0 | 7.0 | 0.0 | 0.0 | 16.9 | 106.8 | 1334.4 |
| Jun | 43.3 | 31.1 | 0.0 | 9.0 | 6.0 | 9.0 | 0.0 | 0.0 | 17.8 | 111.1 | 1388.9 |
| Jul | 42.2 | 30.6 | 0.0 | 11.0 | 6.0 | 7.0 | 0.0 | 0.0 | 17.1 | 107.8 | 1348.1 |
| Aug | 39.4 | 28.9 | 5.0 | 8.0 | 11.0 | 0.0 | 0.0 | 0.0 | 15.3 | 99.1 | 1239.2 |
| Sep | 38.3 | 28.3 | 7.0 | 8.0 | 9.0 | 0.0 | 0.0 | 0.0 | 14.7 | 95.9 | 1198.3 |
| Oct | 37.8 | 26.7 | 9.0 | 6.0 | 9.0 | 0.0 | 0.0 | 0.0 | 13.8 | 91.5 | 1143.9 |
| Nov | 33.9 | 22.8 | 8.0 | 9.0 | 0.0 | 0.0 | 0.0 | 0.0 | 10.7 | 71.3 | 890.7 |
| Dec | 30.0 | 20.0 | 12.0 | 1.0 | 0.0 | 0.0 | 0.0 | 0.0 | 8.0 | 53.3 | 665.8 |
| Ave | 36.3 | 25.4 | 6.5 | 6.7 | 4.5 | 1.9 | 0.0 | 0.0 | 12.7 | 82.6 | 1033.1 |

Figure 4.10   Typical design for solar–diesel hybrid with air conditioning load.

## Customer's requirements

| Customer | Aid Ltd |
|---|---|
| Site | Antofagasta |
| | Lat: 22 S   Long: 69 W   Altitude: 1000 M ASL |
| Load voltage | 24 V DC (nominal); Voltage range 23.1 to 29.7 |
| | Max. VARns: 1% to BATT+2.5% to ARRAY |
| Load consumption | 34560 W Hrs/day (ave.); Equiv. to 1440 W continuous |

## Computed optimum system design

**Solar array:**
Module type   LI1PV (HiE53)
No in series   2
No in parallel   54
Array tilt   30 degrees

**Battery bank:**
Battery type   Lead Low Anty
Cells in series   12 each @ 2.2 V nom
Total capacity   4300 Amp Hrs

**Control parameters:**
Solar: Standard regulation

Diesel: Cut-in at   c. 50% S.o.C.
Cut-out at c. 90% S.o.C.

**Diesel generator set**
Output capacity: 3300 Watts (min)
3600 Watts (ave)
Exercise cycle: 2 Hrs every 9 days

**Design safety factors:**
Array contingency          4.37% corresponding to 5% system oversize
Battery reserve             3 days total

Predicted monthly system performance

| | Radi-ation Lglys | Monthly average figures | | | Ampere hours/day | | | Diesel operation | |
|---|---|---|---|---|---|---|---|---|---|
| | | Generation | | | Load consptn. | Surplus | | Monthly hours | Total starts |
| | | Solar | D.Gen | Batt. | | | | | |
| Jan | 653 | 1182.8 | 147.8 | −17.7 | 1306.1 | 6.7 | | 24.5+6.0 | 2+3 |
| Feb | 602 | 1089.6 | 170.4 | 52.6 | 1305.5 | 7.0 | | 25.8+6.0 | 2+3 |
| Mar | 548 | 991.8 | 370.4 | −54.6 | 1303.4 | 4.3 | | 76.6+0.0 | 6+0 |
| Apr | 542 | 980.1 | 327.6 | 0.0 | 1303.4 | 4.3 | | 65.5+0.0 | 5+0 |
| May | 461 | 831.3 | 428.4 | 45.4 | 1300.9 | 4.3 | | 88.5+0.0 | 7+0 |
| Jun | 402 | 725.2 | 586.7 | −8.8 | 1298.9 | 4.3 | | 117.3+0.0 | 10+0 |
| Jul | 461 | 830.0 | 498.1 | −22.0 | 1301.8 | 4.3 | | 102.9+0.0 | 8+0 |
| Aug | 511 | 920.8 | 376.7 | 9.3 | 1302.5 | 4.3 | | 77.8+0.0 | 6+0 |
| Sep | 548 | 988.0 | 312.8 | 6.0 | 1302.6 | 4.3 | | 62.6+0.0 | 5+0 |
| Oct | 600 | 1083.3 | 258.8 | −30.0 | 1305.4 | 6.7 | | 47.5+6.0 | 4+3 |
| Nov | 621 | 1122.2 | 148.0 | 39.3 | 1302.7 | 6.8 | | 23.6+6.0 | 2+3 |
| Dec | 575 | 1040.2 | 280.4 | −13.2 | 1303.2 | 4.3 | | 61.4+0.0 | 5+0 |
| Ave | 544 | 982 | 326 | 0 | 1303 | 5 | TOT: | 771+24 | 62+12 |

Notes: 1 Fully derated.   2 Negative when there is a net battery recharge.
3 Allowing for voltage effects.   4 Standard operation + exercise time.

@ Antofagasta Meteorological data 22.5 S 68.9 W

| Mth | Horiz Radn | Diffuse % | Temp. | Mth | Horiz Radn | Diffuse % | Temp. |
|---|---|---|---|---|---|---|---|
| Jan | 713 | 11 | 24 | Jul | 356 | 22 | 17 |
| Feb | 609 | 13 | 24 | Aug | 426 | 17 | 17 |
| Mar | 504 | 15 | 23 | Sep | 504 | 15 | 18 |
| Apr | 452 | 17 | 21 | Oct | 609 | 12 | 19 |
| May | 356 | 22 | 19 | Nov | 678 | 11 | 21 |
| Jun | 304 | 28 | 18 | Dec | 643 | 12 | 22 |

Figure 4.9   Solapak Limited hybrid system design report.

thus be assessed and the impact on solar array size and diesel run time balanced to best effect.

### 4.6.5 The Difficult System

The "difficult" system incorporates two power generators, both of which use variable input conditions.

#### 4.6.5.1 Solar-Wind Hybrid

Such systems have been installed in a variety of countries and with very variable results. Where failures have occurred, this seems usually to have been due to the wind turbine or to the battery system. However, recent advances in wind turbine technology, especially in relation to the blades, and the development of systems design integration procedures now yield systems capable of commercialisation.

Assuming proper engineering standards, the practical difficulties for wind turbines have now largely been overcome, at least for the higher power generators. The design algorithms required for systems with two varying inputs demand higher safety margins than would otherwise be applied to, say, a solar-TEG or solar-diesel system. By definition, this leads to large system surpluses being present before the prospective user has the confidence to run a system commercially. Figure 4.11 gives a graphical representation of a solar-wind system typical of the UK. An example of this type of system is discussed in more detail in Section 4.7 of this chapter. From Figure 4.11, the reader can compare the surpluses embodied in this design with the solar-diesel system.

### 4.6.6 Other Hybrids

Cogeneration systems comprising three or four elements can be envisaged. There are others which must already exist, ie. solar photovoltaic/thermal, which have not been included and others still to be found (PV-biomass or PV-solar pond?). However, it is the solar-diesel hybrid, with its combination of two reliable, controllable and maintainable generators that has the potential to make the most impact in this world of photovoltaic hybrid systems, in the immediate future at least.

### 4.7 A Practical Hybrid Application

As an example of a practical hybrid application, the solar-wind system installed by Solapak at Milton Keynes in 1986 will be discussed. The system uses power derived from a solar array and wind turbine to feed nine residential houses (Figure 4.12) installed on a site known as the Energy Park. This development area has been specifically targeted for the demonstration of housing stock with low energy needs and,

Figure 4.11   Typical characteristics of a solar–wind hybrid (© 1987 Solapak Ltd).

### Basis of design
Solar and wind systems designed so that combined output meets demand in all seasons
(even at minimum expected wind levels).

Battery bank to absorb hourly and daily fluctuations
surplus power utilised elsewhere.

### Primary design parameters
Solar array:          4.1 kWp
W.D.G. capacity:  60 kW maximum
Battery storage:    140 kWh maximum
Inverter capacity: 25 kW minimum

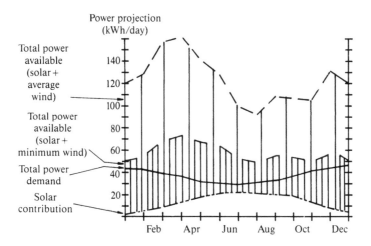

Predicted performance

| | | | | | | | | | | | | |
|---|---|---|---|---|---|---|---|---|---|---|---|---|
| Demand (kWh/day) | 43 | 39 | 36 | 31 | 29 | 28 | 29 | 31 | 35 | 39 | 42 | 44 |
| Solar (kWh/day) | 5 | 9 | 13 | 17 | 20 | 21 | 19 | 18 | 16 | 11 | 5 | 2 |
| Min wind (kWh/day) | 49 | 59 | 59 | 49 | 43 | 31 | 28 | 35 | 35 | 37 | 49 | 47 |
| Surplus (kWh/day) | 11 | 29 | 37 | 36 | 34 | 24 | 19 | 22 | 17 | 9 | 12 | 5 |
| Sun (Hrs) | 1.7 | 2.5 | 3.7 | 4.7 | 5.3 | 5.5 | 5.1 | 4.9 | 4.4 | 3 | 1.7 | 1.1 |
| Wind speed | 5.6 | 5.9 | 5.9 | 5.6 | 5.3 | 4.8 | 4.6 | 5 | 5 | 5.1 | 5.6 | 5.5 |

against this background of architectural innovation, it is very appropriate to consider the alternative forms of power generation which, in the context of placement in an urban environment, need to be as unobtrusive as possible, both visually and electrically.

In the UK, wind availability in winter can be up to five times higher than in summer, whilst solar availability can be eight times higher in summer than in winter. With the average winter energy demand being about double the summer demand, a mix of complementary co-generation devices should be able to meet the required summer/winter profile.

The scheme is configured so that the 4.6 kW solar array and the 22 kW wind turbine feed into the Energy Management Centre where the energy is then distributed to the various main loads:

- AC power to the 9 houses from an inverter. System sizing has been established to meet a proportion of the total housing energy demand so AC power export to the grid from the scheme has not been facilitated.

- AC heating load in each of the 9 houses direct from the wind turbine

and to the "intermediate" loads:

- Battery charging from solar and wind turbine

- AC dump load selectively applied to the wind turbine at times of no housing demand and with battery full

The aims of the project are to:

- show how solar and wind regimes can provide complementary generating capacities without interseasonal storage

- implement a scheme that provides AC power to the housing load in such a way as the 'source' (solar/wind or grid) is transparent to the householder

- integrate this new technology into a mainland urban environment

- evaluate the availability and utilisation of the scheme

- examine and identify improvements to system and sub-system control mechanisms

- determine how similar principles can be adopted, with different summer/winter, solar/wind profiles in other areas of the world.

The control logic gives priority to the solar/wind system at all times. In normal operation, housing load is transferred to the mains grid only when the inverter rating is exceeded or when the battery is discharged.

The nine houses were completed and occupied by June 1987 since when demonstration and engineering trials have been undertaken and the monitoring system validated.

### 4.7.1 Solar Array

In addition to the arrays on each of the nine houses, an additional 2.6 kW solar array is mounted on the roof of the Energy Management Centre which has been designed to represent a double garage complete with pitched tiled roof (Figure 4.13). The roof modules are assembled onto horizontal tubular supports which span the width of the building (orientated E-W) and fit into standard industrial bearings at each end. Five such assemblies are mounted on two steel bearer channels supported by brick piers at the four corners of the building. This has proven to be a very effective way of roof mounting arrays without jeopardizing the roof waterproofing, and at the same time allowing for easy seasonal tilt adjustment.

### 4.7.2 Solar Controls

The modules in each of the houses are wired in series and each house is cabled back to the EMC separately to a control unit which provides the following:

| | |
|---|---|
| Regulator switch | - 14 off series FET with blocking diodes<br>- opto-isolator on each switch for voltage isolation from control drivers |
| Regulator control | - cascade operation<br>- separate battery voltage and temperature sensing |
| Battery monitor | - high volts signal<br>- low volts signal<br>- load cut signal<br>- interface to power conditioning |

### 4.7.3 Wind Turbine Generator (WTG)

| | |
|---|---|
| Configuration | : upwind horizontal axis machine with AC generator |
| Rating | : 22 kW at 12 m/s |
| Rotor Diameter | : 6 m |
| Hub Height | : 18 m |
| Tower | : tubular 2 section tower |
| Safety Mechanisms | : - wingtip air brakes |

Figure 4.12     View of one of the houses in the Milton
                Keynes Energy Park. The photovoltaic panels
                are installed in the conservatory roof.

Figure 4.13     The Energy Management Centre, which has a
                2.6kW solar array mounted on the roof.

|                    |   | - shaft mounted disc brakes released by compressed air |
|--------------------|---|--------------------------------------------------------|

Control Mechanism : - main rotor turned into wind by pair separate "fantail" rotors

- shaft mounted disc brakes released by compressed air
- 10 automatic internal status checks including vibration, cable twist, overspeed
- main rotor turned into wind by pair separate "fantail" rotors
- start up by wind speed detection. Power start provided for this location initiated at 6 m/s wind speed
- rotation speed governed by automatic multi-stage control pickup or dumping electrical load.

The power start feature has increased the power output from the WTG by approximately 24% with measurements made so far. This is expected to increase towards the predicted 30% during the evaluation period, subject to wind speeds experienced. The wind turbine is shown in Figure 4.14.

### 4.7.4 Power Conditioning Unit, Back Up Power Supply and Operating Strategy

The nature of the users in this scheme meant that breaks in supply when changing between the solar/wind scheme and mains grid, or vice versa, could not be tolerated nor could variations in supply frequency be accepted. In fact, it was considered important that as far as the user was concerned, the source of power should be as transparent as possible. This led to the choice of an Uninterruptable Power Supply (UPS) system as offering the most satisfactory solution, where a static switch is used to perform the no brake changeover function between the solar/wind scheme and the mains grid. The UPS system also incorporates a 3 stage 30 A battery charger which can be fed either from the wind turbine or from the mains grid by operation of an external manual switch.

System supply priorities are as follows:

- battery is charged from solar array
- wind turbine feeds power in priority according to the instantaneous output power
    - into the battery via the three stage battery charger
    - into the inverter
    - into the heating load
    - into the AC dump load
- inverter feeds AC power to the housing load.

No wind turbine output - inverter runs until battery partially depleted.

Figure 4.14   The 22kW wind turbine installed at Milton Keynes.

Grid failure - inverter starts and runs until grid restores or battery fully depleted.

Housing load transferred to grid - when inverter shut down
                                     - inverter overload.

This strategy is intended to find an optimum match of energies from solar and wind, and provide good supply efficiency and safety. It is intended to give both good systems availability and energy utilisation.

During normal operation the inverter output to the housing load is synchronised to the grid frequency. If the grid fails, the inverter will continue to run with the frequency determined by an internal crystal. If the battery becomes depleted or the load is in excess of the inverter rating (other than short term overload), then the bypass switch will transfer the housing load to the grid supply without break. A separate manual switch allows the housing load to be transferred to the grid supply in case of emergency. This has not yet been used.

## 4.8 Conclusions

This chapter has discussed photovoltaic systems for professional applications, including the use of hybrid systems. The most commercially successful applications are those in telecommunications and cathodic protection, where PV systems are often first choice for both technical and economic reasons, although, in some cases, backup from diesel or other sources is to be recommended. There is a growing market for systems designed for water treatment, water pumping, ventilation, refrigeration etc. and the possibility of using a combination of sources in a hybrid systems expands the number of cost effective applications. The use of hybrid systems is particularly attractive for provision of the variety of electrical needs in dwellings or public buildings. It is to be expected that the market for PV and PV hybrid systems for professional applications will continue to expand.

## 4.9 References

4.1 The Solamax controller is a maximum power point tracking unit available from Solapak Ltd.
4.2 Telecommunications Journal, November 1983
4.3 35,000 kWh Later - A User's Experience of Photovoltaics, G.S.M.Teale, Petroleum Development, Oman.

# Low Power Applications of Photovoltaics

Nicola M. Pearsall and Robert Hill
Newcastle Photovoltaics Applications Centre
Newcastle upon Tyne Polytechnic
Newcastle upon Tyne, NE1 8ST

## 5.1 Introduction

One of the often cited advantages of a photovoltaic (PV) power system is its modularity. This allows the system to be designed to meet a wide variety of load requirements and to be expanded easily if the electricity demand increases. Whilst this is clearly useful for the applications discussed in previous chapters, it is, perhaps, the most important aspect of photovoltaics when low power applications are considered. Although they rarely attract the publicity of the large PV installation or the latest satellite, this type of application is one of the most commercially successful to date, due to a combination of versatility, ingenuity and shrewd marketing policy.

For the purposes of this chapter, low power applications have been defined as those with an installed power capacity of less than that provided by a standard module (about 40W) and, therefore, can be considered as a custom designed product. Some of the systems discussed may be available with higher rated arrays (i.e. several modules) but with no change in design principle. The lower limit of PV array size is, of course, a single cell. Since the array size is both small and non-standard in configuration, it is generally designed specifically for the application in question. This is in contrast to the general case for the larger installations, where multiples of standard 40-50 W modules are used.

In other chapters, the large scale use of photovoltaics to replace conventional energy supplies, such as diesel engines, is discussed. Clearly, the low power PV array is used as a replacement for another low power electricity supply. Most commonly, this is some form of battery, but can also be mains electricity. The systems can be divided into two kinds: consumer products, in which the PV array is an integral part of the product, thus affecting design criteria, and power supply products, where the PV array itself is marketed and acts as a small replacement power supply. The latter category is similar to the larger scale installations discussed in other chapters, although there may be different constraints on construction, mounting and cost of the array. Examples of both types of system will be discussed in more detail later in this chapter.

Although many kinds of solar cell are under investigation in the research laboratories, only crystalline silicon has been used in large scale terrestrial installations. Due to the requirement for assured long term reliability, it seems likely that crystalline silicon will continue to be the most common type of module for remote applications for some years to come, although thin film modules are expected to be used increasingly. In contrast, the consumer market is much more heavily reliant on thin film amorphous silicon cells than on crystalline silicon cells, due to their adaptability in terms of mounting, their performance at low light levels and their relative versatility in array sizing (Figure 5.1). The use of amorphous silicon cells will be discussed in the section on consumer products.

## 5.2 Conversion of Light to Electricity

The detailed design of low power PV arrays, especially in cases where there are constraints on the size, access to light etc., relies on a knowledge of the cell performance under the relevant environmental conditions. To enable discussion of array design and an appreciation of the design features of these consumer products, the basic principles of cell operation will be reviewed. Particular emphasis is placed on the difference in performance between crystalline and amorphous silicon cells, since these are the two cell types most commonly used at present.

There are many different structures for commercial solar cells, especially in the case of amorphous silicon, where there is more scope for innovative cell design. It would be inappropriate to discuss these structures in detail in a publication of this type. Rather, the basic operational parameters, common to all devices, will be reviewed. Readers are referred to publications regarding particular cell structures for further information [Ref. 5.1, 5.2].

### 5.2.1 Output Parameters

Solar cells are semiconductor diodes in which the rectifying junction can be formed by different impurities in a homogeneous material (homojunction), by the interface of two different semiconductors (heterojunction) or by the interface of a metal with a semiconductor surface (MS or MIS junction). For a given illumination (defined in terms of both intensity and spectral content) and at a given temperature, the output of a cell can be defined in terms of three parameters. These are the open circuit voltage, $V_{oc}$, (that is, the voltage when the load resistance is infinite), the short circuit current, $J_{sc}$, (that is, the current when the load resistance is zero) and the fill factor, FF,. The current-voltage characteristic of a solar cell under illumination is shown in Figure 5.2, illustrating the short circuit current, open circuit voltage and the maximum power

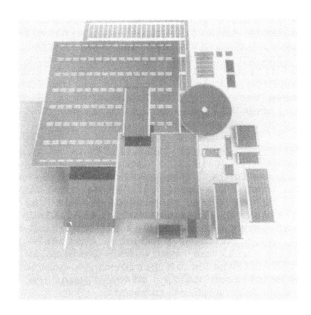

Figure 5.1        Amorphous silicon modules of various shapes
                  and sizes for use in a variety of products.
                  (Photograph by courtesy of Arco Solar Europe
                  Ltd.)

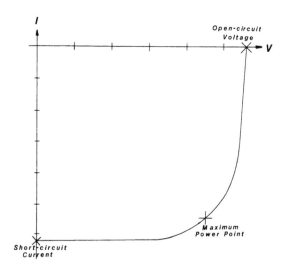

Figure 5.2        Current-voltage characteristic of a solar
                  cell.

point, the point on the curve where the product of the
current and voltage is at a maximum. The fill factor is
defined as the ratio of the power at the maximum power point
to the product of the open circuit voltage and the short
circuit current. The maximum power is obtained from the cell
if the load is of a value such that the cell is operating at
the maximum power point. The efficiency of the solar cell is
equal to the ratio of the maximum power obtainable to the
power input (from the light). It can also be defined in
terms of the three cell parameters as follows:

Efficiency $= V_{oc}$ x $J_{sc}$ x FF / Power in

For a typical single crystal silicon cell, $V_{oc}$ is about 0.6
V, $J_{sc}$ is 25-30 mA/cm$^2$ and FF is about 0.75-0.80, in
sunlight of intensity 100 mW/cm$^2$ and at a cell temperature
of 25$^o$C. For a typical amorphous silicon cell, these values
would be $V_{oc}$ of about 0.85V, $J_{sc}$ of about 15 mA/cm$^2$ and FF
of about 0.65 for the same test conditions. Thus, the
efficiency of this cell is somewhat less than for a
crystalline silicon cell, but it has a higher $V_{oc}$, which can
be useful for some applications. The current is linearly
proportional to the intensity of the incident light and,
clearly, the current available from a single cell increases
linearly with increasing cell area. The voltage, however, is
independent of cell area and increases logarithmically with
increasing incident light intensity.

The open circuit voltage which can be generated by a solar
cell at a given light intensity depends on the materials
from which the cell is made, the electronic design of the
junction and the temperature of the cell.

Electronic devices usually operate at voltages greater than
that which can be obtained from any single solar cell, so it
is often necessary to connect a number of cells in series to
provide the voltage required (Figure 5.3). For consumer
products, the cost can be reduced by minimising the the
number of cells needed to give the required voltage, so
cells with a large operating voltage are advantageous. The
current output from a cell can be increased by increasing
the area of the cell, either by making each cell larger or
by connecting cells of the same size in parallel.

5.2.2 Spectral Response

The current available for a single cell, of any type,
depends on both the intensity and the spectral content of
the incident light. This is a consequence of the fact that
the cell does not respond equally well to all wavelengths of
light. The spectral response of the cell depends on the
materials from which it has been fabricated and the
fabrication method and conditions. The spectral response
curves of typical crystalline and amorphous silicon solar

cells are shown in Figure 5.4a. It can be seen that both the
shape and the wavelength at which the response is maximum
vary and the variation depends on both the cell material and
the detail of the cell structure. Since different light
sources vary in spectral content, as illustrated in Figure
5.4b, it is clear that some cells will be more suited than
others to operation under a particular source. For white
light sources, whose energy is spread over the visible and
infra-red wavelengths, the overall response is an average
over the wavelength range of the light and, thus, is much
lower than the peak response. If monochromatic light is
used, however, the wavelength can be chosen to coincide with
the peak response of the cell and the spectral output of the
light source be matched as closely as possible.

### 5.2.3 Low Illumination Conditions

PV power supplies in consumer products are often used in
conditions of domestic or office room lighting rather than
in sunlight and the intensity of light falling on the PV
array may be only one hundredth or less of the intensity of
bright sunlight (Figure 5.5). Thus, the behaviour of solar
cells under low illumination is a very important factor in
the choice of cells for consumer products. As stated above,
the current output of a cell varies linearly with incident
light intensity for bright sunlight, but this simple
relationship does not always apply at low light levels
(below about 10 mW/cm$^2$). The voltage output of an ideal cell
will vary logarithmically with light intensity, so, for
instance, a cell with $V_{oc}$ of 590 mV in strong sunlight would
have $V_{oc}$ of 530 mV at 1/10 intensity and 470 mV at 1/100
intensity. Crystalline or multicrystalline silicon cells
lose voltage much more rapidly than the ideal cell when the
light intensity falls below about 1/10 bright sunlight and
at low light levels may be unable to generate the voltage
required to sustain operation of the circuitry. Thin film
cells, such as amorphous silicon, maintain voltage down to
much lower light levels and a calculator, for example, with
amorphous silicon cells will continue to operate even in
quite dim lighting.

### 5.2.4 Stability

The final aspect of cell behaviour to be discussed here is
that of cell stability. The cell must continue to give the
required output after operation under appropriate conditions
for the duration of the application being considered. The
operating conditions can be in sunlight or, in the case of
some consumer products, predominantly room light, and the
temperature variations experienced by the cell will depend
on the location of that cell. The duration of the
application varies from 10-20 years, for some large scale
terrestrial installations, to 1-2 years, for consumer
products such as calculators. In the early stages of

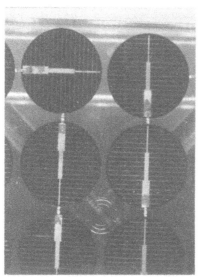

Figure 5.3     Section of crystalline silicon module showing the interconnection of the cells. The cells are series connected, i.e. the front of one cell is connected to the rear of the next cell in the string.

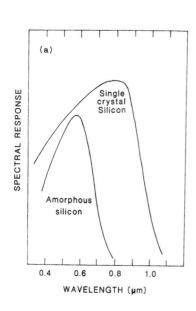

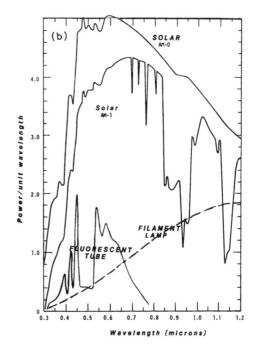

Figure 5.4     a) Typical spectral response curves for different types of silicon solar cell; b) Spectral content of different light sources.

development, amorphous silicon cells exhibited serious instability under high illumination conditions, making them unsuitable for use in full sunlight. Improvements in stability have been made and the cells now exhibit lifetimes well in excess of those required for consumer product applications. Their suitability for long term, large scale installations has yet to be proven, although the stability of new designs of amorphous silicon cells and modules has been greatly improved. The crystalline silicon cell displays a high degree of stability and encapsulation techniques are such that modules now have proven lifetimes (mean time between failures) of well over 10 years.

5.3 <u>Consumer Products</u>

We have defined consumer products as those in which the photovoltaic power source is an integral part of a product, which is marketed as a total commodity. Thus, the market is not that of a power supply, but of a calculator, fan, watch etc. The product is sold on the basis of its performance, convenience, attractiveness and cost in comparison to other similar products, whether PV powered or not. Thus, the PV component must be capable of performing its function without impeding the operation of the product, from either a practical or aesthetic viewpoint, and should have either operational or cost advantages over an alternative power supply.

When considering the incorporation of a PV power source in any product, the following questions must be asked:

(i) Is a PV array practical?

(ii) Is a PV array cost-effective?

Clearly, in many cases, such as the examples of calculators and watches, the answers to both these questions are in the affirmative. We will consider the requirements for this to be the case.

5.3.1. <u>Types of Load</u>

For the design of all PV systems, whether large or small, the first consideration is the size and nature of the load. Even where the sunlight is not the principal light source, there is a variation in the light incident on the array and, therefore, the array must be sized so as to give the required output under all typical conditions of illumination. Two types of load, giving different design requirements, can be defined:-

(1) A load which is present only when light is available (e.g. calculator)

(2) A continuous load, or one which is present at periods of no illumination, leading to the need for storage of electricity (e.g. clock)

Clearly, the first case is the easiest to consider, since the array is sized simply to provide sufficient power to run the unit under the prevailing light conditions.

### 5.3.2 Non-Storage Product

As an example, we will discuss one of the first, and most successful, applications of PV to a consumer product, the pocket calculator (Figure 5.6). The load for the system is well defined, being the circuitry and display unit. Progress in the field of electronics has led to a reduction in the power requirement over recent years, thus reducing the size of the PV system required. For economic viability, it is advisable to ensure maximum interchangeability of components, such as the power supply system, between different models within a company's product line. Thus, most circuitry will have been designed to operate from the same power supply system, for example, 3 x 1.5V batteries. The versatility of the PV array allows that battery system to be emulated, i.e. an array giving 4.5V to be constructed, leading to the use of common circuitry for both battery and PV powered models. Fortunately for both PV and small battery power systems, the current required by calculator circuitry is small.

The number of series connected cells needed to provide the required voltage depends on the type of cell used. As discussed previously, since each amorphous silicon cell will give a higher voltage output than a crystalline silicon cell, fewer amorphous silicon cells are required to produce a given voltage. One consequence of this is that there are fewer cell interconnections for a given voltage rating of an amorphous silicon array. The current requirement dictates the area of each cell and, hence, the area of the total array.

The sizing must be carried out with reference to the likely light input conditions. A calculator does not require storage of electricity, since it is most unlikely to be used in darkness. However, the pocket calculator is most often used in an office environment, that is, under illumination from an artificial light source. In fact, the arrays used are often considerably oversized for use in sunlight, in order to allow for normal operation in low light level conditions. In the previous section, the response of cells to different light sources was discussed and it was clear that the amorphous silicon cell performs better than a crystalline silicon cell under fluorescent light conditions. In addition, it was explained that the amorphous silicon cell also shows superior voltage performance under low light

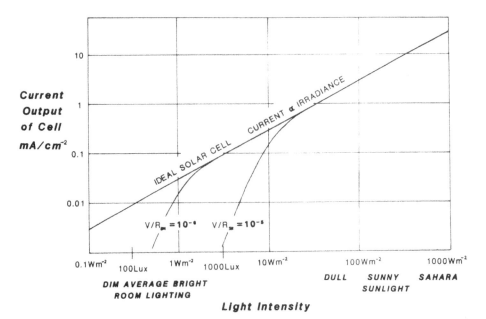

Figure 5.5    Graph of cell current against light
intensity, for a range of lighting
conditions. A crystalline silicon cell is
assumed.

Figure 5.6    Hand held pocket calculator, powered by four
amorphous silicon cells, which can be seen at
the top of the calculator.

levels. This performance advantage is one of the reasons for the widespread use of amorphous silicon cells in consumer products, even though their efficiency under standard sunlight conditions is less than that for crystalline cells.

Having established the required size of the PV array, the product design must be both practical and attractive. Maintaining the same example, the pocket calculator, the usual design is to place the PV component above or alongside the display unit, since this is the position least likely to be shadowed during operation of the keys. A PV array placed below or amongst the keys would be likely to perform poorly due to intermittent shadowing. The required array size must be in proportion to the rest of the product in order to be visually pleasing. For example, a product is unlikely to be successful in the market place if it must be doubled in size to accommodate the PV power supply or if the need to ensure illumination of the array interferes with the ease of normal operation.

The previous discussion has addressed the first criterion for the adoption of a PV powered product, that of its technical and practical feasibility. However, of crucial importance is the question of the economic feasibility of the product. The relative costs of the power supply options must be considered. This topic includes the differences in the manufacturing processes for the various designs, as well as the power systems themselves. For example, the use of a battery power system for a pocket calculator requires the provision of a battery compartment. This entails extra pressings in order to provide a removable cover and may require extra strengthening of the calculator body. By contrast, the PV powered calculator body may be constructed without such a compartment. The extra cost of providing the battery compartment must be included in the costings for the battery power supply. On the other hand, it is usual for the calculator manufacturer to sell the product without a battery provided. This is clearly not the case for the PV powered calculator, where the power supply is an integral part. Thus, the extra cost of the battery may not be included in the product costings.

The use of a PV power supply also allows the manufacturers to develop products, such as the "credit card" calculator (Figure 5.7), which would be difficult and expensive to produce in a battery powered version. Thus, the use of PV power provides an opportunity to develop novel products which create new markets and, therefore, it may be commercially advantageous even if the cost difference between PV and batteries is small for existing products.

For the customer, the cost comparison is between electricity from photovoltaics and from small batteries. Since these small batteries exhibit a high cost per Watt, as shown in

Table 5.1, this is a very different condition than for competitiveness with large scale electricity supply, as discussed in the other chapters. The manufacturer will often charge a somewhat higher price for the PV powered product. The extra amount which can be sustained, whilst maintaining a higher sales quota than for the non-PV powered unit, depends on the cost to the consumer of the replacement batteries and the perceived value of the convenience of the PV powered unit.

The lifetime of the power supply is rarely the limiting factor in determining the lifetime of the product. Many calculators are replaced, due to malfunction, loss or the desire for a more recent model, within two to three years of purchase. Thus, the lifetime of the power supply must be sufficient only to outlast the rest of the product. This is a much easier condition to achieve than the long lifetimes (10-20 years) required for large scale installations.

### 5.3.3 Products with Storage

The second loading condition is that in which the electrical power is required at times when there is no illumination. In this case, some form of electrical storage is used. Until recently, this has been a rechargeable battery, of capacity consistent with the electricity demand.  Several items, such as clocks (Figure 5.8), now have a capacitor storage system, which has become possible due to the development of physically small, high storage capacitors. The economic, practical and aesthetic considerations of the use of PV cells in applications both with and without storage are very similar.

### 5.3.4 Marketing of the Product

Finally, we must consider the marketing of the product and the points on which the PV powered product may be sold. Clearly the first advantage is the portability of the product, which, given adequate illumination, is always ready for use.  Thus, not only does the user escape the need to purchase replacement batteries, they also avoid the situation of failing or failed batteries at a time when replacement is not possible.  In addition, the product does not rely on the provision of mains electricity at any time. Thus, the main selling point for a PV powered product is its convenience.  Any design feature which may affect that convenience of use adversely must be avoided, if possible. Lastly, the use of solar power is fashionable at present and, thus, the product marketing may emphasise this aspect. This factor has a limited lifespan and the fashionability of using PV power should not be allowed to outweigh practical or economic considerations during the design of a product.

It is worth noting the different conditions of introduction of PV into the consumer product area, when compared to the introduction into the large scale power supply field. The cost of a product is partially dependent upon the size of the market for that product, due to the cost benefits of large scale production and the financial optimism associated with a successful commodity. The price of standard terrestrial PV modules, although significantly reduced over the last ten years, has not decreased as fast as was hoped, due to a relatively low market penetration, and, of course, the growth of the market is restricted by the commodity price. Some progress has been made, due to third party financing and extensive tax credits in the USA and to more aggressive marketing by the major PV manufacturers, and cost reductions are still forecast with confidence.

By contrast, the market for PV arrays in consumer products was large enough to allow considerable investment at the beginning. This was possible because of the vertically integrated structure of the Japanese electronics companies. They produced the cells themselves and, with a guaranteed market in products manufactured by their own company, they soon achieved costs which proved economically competitive with other types of power supply. This technique was applied successfully to the production of PV powered calculators and watches, before expansion into other types of consumer product.

The success of these first products and the versatility of the PV power supply has led to a wide range of consumer products, including calculators, watches, clocks, radios and other audio equipment, hats incorporating cooling fans or radios, toys, kitchen and garden tools, pumps for solar hot water systems, small battery chargers etc (Figures 5.9-5.11). The variety of products which can incorporate PV power supplies is very large and should grow considerably over the next few years.

## 5.4 Replacement Power Supplies

This section discusses the use of small modules or small arrays as a replacement power supply for another form of low power supply or as a means of battery charging. Here, we are considering photovoltaics alone. Hybrid systems (e.g. wind/PV, diesel/PV) are considered in Chapter 4 in this volume and are more suited to the larger power demand applications. The constraints on the use of photovoltaic modules for replacement power supplies are similar for both small and large power requirements, being the applicability of using solar cells, the availability of light and the relative costs of the power supplies. For low power applications, where the system may consist of single modules only, there may be constraints of special mounting conditions, size etc.

Table 5.1        Cost of electricity from primary batteries

| BATTERY TYPE | COST (£/KWHr) |
|---|---|
| Button Cell | 10,000 |
| AA Cell     (Zn/C) | 800 |
| D Cell     (Zn/C) | 500 |
| D Cell     (Mn/Alk) | 300 |
| Lantern Battery (Zn/C) | 200 |

Figure 5.7        The "credit card" calculator, powered by
amorphous silicon cells mounted alongside the
display.

Figure 5.8     Solar powered clocks. For the larger versions
               at the rear of the photograph, the PV module
               forms the face of the clock. (Photograph by
               courtesy of Chronar UK Ltd.)

Figure 5.9 Solar powered radio mounted on a cap.

Figure 5.10     This model windmill, powered by the small
                solar panel on the right, is typical of the
                solar powered toys available.

Figure 5.11     Battery charger for use with 1.5V nickel-
                cadmium rechargeable batteries. The unit is
                powered by the small amorphous silicon
                module. (Photograph by courtesy of Chronar UK
                Ltd.)

The requirement for a successful photovoltaic implementation of the kind considered here is a demand for a small amount of DC electricity. This may be supplied at present by a battery, by an alternative power supply, such as a diesel engine, or, in some cases, not supplied at all due to difficulty or cost. These applications often require storage, to allow use of electricity outside sunlight hours, and the usual configuration is to use the photovoltaic array to charge storage batteries. Thus, photovoltaics can be used in locations where, on average, there is sufficient sunlight during a 24 hour period to provide electricity to be used in that period, allowing for the efficiency of the total system. These are effectively the same conditions as for larger installations, but specialised applications may require non-standard modules.

At present, there are two major markets for small PV power supplies, these being the leisure industry and military applications. The former is the most diverse and, therefore, interesting to discuss. The term "leisure industry" covers many aspects, including holiday homes, boating, motoring, DIY, gardening, etc. This wide variety of applications calls for different types of module depending on the conditions which the power supply must fulful. The military applications are mainly battery charging and power supply for telecommunications equipment.

It is worth examining the advantages of the PV power supply which make it popular in these two widely different areas. The general public at leisure have one main factor in common with the typical soldier. Both have a tendancy to find themselves in locations remote from the normal supplies of electricity, whether on a military exercise, motoring in the countryside, sailing the oceans or, simply, at the bottom of the garden. Photovoltaics offers a power supply which is quiet, reliable and, importantly, portable. Additionally, whilst it is always inconvenient to discover that your battery is flat, in some cases, such as loss of communications or navigation aids, it can be very dangerous. There is considerable value in having a means of ensuring that your battery is always in an operable condition. The ease of maintenance and operation of a PV module is very attractive to the leisure customer, since alternative power supplies, such as diesel engines, often require both regular maintenance and fuel.

Since there are so many different applications, requiring different module designs, it would not be practical to consider them all here. Rather, a few examples, illustrated by the accompanying photographs (Figures 5.12-5.17), will be discussed to show the variety possible. For PV modules for military use, the requirements are a high degree of portability, by backpack or rough terrain vehicle, and rugged construction.   The panel, as illustrated in (Figure 5.12), is usually designed to be folded for easy carrying, being deployed when required.

Portable power packs are also available for powering a large number of non-military applications, from radios and television to instrumentation (Figure 5.13). The size and configuration of these power packs varies according to the application (or range of applications) for which they are intended, but they are all designed to provide an easy to use, portable power supply.

PV panels for boats, on the other hand, are not required to be portable since they are usually fixed in place.   Ideally, they should be deck mountable.   The modules do, however, have to be resistant to both the marine environment and the rigours of shipboard life.   The illustration (Figure 5.14) shows a module in which the cells are encapsulated in a polycarbonate material, which makes the module very resistant to impact damage, even from crewmen walking on it. Other manufacturers produce glass encapsulated modules for operation in marine conditions.

Figure 5.15 is a photograph of a car sunroof, incorporating amorphous silicon cells and designed to supplement the electrical supply of the vehicle.   This application clearly imposes constraints on the size, geometry and appearance of the module.   Amorphous silicon cells are popular in this context, since, as can be observed from the photograph, they allow some of the light to pass through the sunroof of the car.   At the time of writing, these sunroofs were at the early stages of marketing and were not widely available. More common with respect to motor vehicles is the use of small panels placed on the dashboard (Figure 5.16) or mounted on the window (Figure 5.17), which power a small fan to reduce the interior temperature during parking in hot, sunny weather or which keep the battery charged during long parked periods in an outdoor carpark, such as at an airport.

The usual configuration for a small PV power supply is a module(s), optional controller and battery pack, if storage is required.   There are several different types of batteries which may be used and the choice of battery depends on the specific application.   The battery must, however, be capable of being trickle charged from the PV array, it must withstand occasional deep discharges without damage and have both high storage efficiency and a lifetime of some hundreds of cycles.   No one type of battery is optimum for all applications and the choice will depend on the relative importance of the above characteristics and the battery cost.

The use of PV modules for low power applications is becoming more widespread and will continue to do so as modules are

Figure 5.12    Folding panel for portability. (Photograph by courtesy of PAG Solar Technology Ltd.)

Figure 5.13    Portable power pack, incorporating folding panel. (Photograph by courtesy of AEG (UK) Ltd.)

Figure 5.14    PV panel mounted on yacht deck. (Photograph by courtesy of PAG Solar Technology Ltd.)

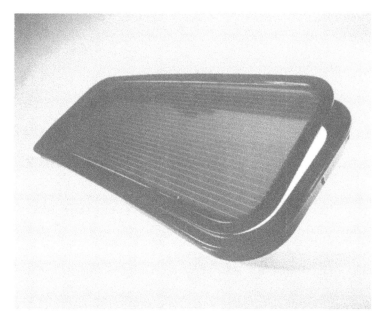

Figure 5.15          Amorphous silicon module incorporated into
                     car sunroof, for supplementing the electrical
                     supply of the vehicle. (Photograph by
                     courtesy of Arco Solar Europe Ltd.)

Figure 5.16          Solar battery charger for mounting on car
                     dashboard. (Photograph by courtesy of Chronar
                     UK Ltd.)

introduced at a more accessible price and as the consumer becomes familiar with the possibilities of this power supply. The market has improved with the introduction of small modules, most notably based on amorphous silicon, which, whilst at similar price per Watt of output to the larger modules of the same type, are of a price per unit which can be afforded. It is important to recognise that the customer is purchasing a service such as "cool car", "fully charged battery" or "operating radio" and is concerned with the cost of the service rather than the cost/Watt of the modules. It has become possible for the customer to purchase a PV module from some hardware stores at a similar unit price to the other equipment on sale and, as this becomes more widespread, the use of PV equipment will become more common.

## 5.5 Instrumentation

It has been emphasised already that photovoltaic cells can provide electrical power at the point of use, without the need for transmission over significant distances. This factor makes the technology very attractive for use in remote locations away from the utility grid and can also be important for the provision of small amounts of electrical power to instrumentation in difficult environments.

Many of the sensors and other measuring equipment vital to the safe and efficient operation of modern industry can be powered by a small D.C. electricity source. Traditionally, this has been a battery, but the desired location of some of these sensors mean that it can be extremely difficult, if not impossible, to replace or recharge these batteries at regular intervals. Power must be transmitted to the sensors, therefore, usually by electrical cabling. This may be unacceptable when the cabling must pass through a hazardous environment, e.g. where there is a risk of explosion.

The photovoltaic alternative is to transmit the power in the form of a light beam and to convert the light to D.C. electricity at the sensor itself. The light may be in the form of either a free or guided beam, depending on the application, and, if desired, a monochromatic light source may be used, in order to increase cell efficiency. With specially designed cell mounting and encapsulation, the system can withstand many environments.

This application is still in the very early stages, although the use of PV power supplies for monitoring of pipelines etc. is common in locations with appropriate climates. Although use in industrial sensors is unlikely to result in a large market for photovoltaics (in terms of the power produced), it will provide outlets for fibre optic cabling, light sources etc. and, most importantly, it could help to solve difficult monitoring problems for industry.

## 5.6 The Low Power Applications Market

Although some small consumer products, such as toys, have
been produced since the early days of photovoltaics, the
upswing in the low power market, particularly in consumer
products, came with the development of the amorphous silicon
cell. The advantages of the a-Si cell for this kind of
application have already been discussed.

The first a-Si photovoltaic devices were made at RCA
Laboratories in 1974 [Ref. 5.3] and have been the subject of
intense development effort since then. However, it was only
in 1980 that Sanyo Electric Co. began to sell pocket
calculators powered by a-Si:H cells [Ref. 5.4]. Other
manufacturers, mainly Japanese, followed their lead and the
market grew so rapidly that, in 1984, Japan shipped
approximately 100 million solar powered calculators. Sanyo
alone produced 3.2 MW of a-Si cells for consumer products
and, in 1985, approximately doubled their capability to the
level of around 8 million calculators per month [Ref. 5.5].
This rate of growth could not continue, of course, and the
PV powered calculator market appears to have levelled out at
about 130 million units per year, now that it has become an
expected part of the calculator product line. The trend in
market growth seems likely to be repeated by PV powered
analogue watches and clocks, using the new range of high
performance capacitors.

Not surprisingly, the worldwide market figures for
photovoltaics reflected the upward trend in solar powered
consumer products. Figures for 1983-87 are presented in
Table 5.2. Amorphous silicon devices accounted for about 14%
of the total shipped in 1983 (in terms of power rating),
whereas this share had grown to almost 30% in 1985. Since
then, the market share has stayed almost steady, although
the amount of product sold has risen somewhat.

The data in Table 5.2 are presented in MW of product sold.
This can be a rather misleading practice when considering
consumer products. The figures are calculated by multiplying
the area of devices produced by a typical efficiency value,
determined under standard sunlight conditions. Since these
cells are, in the main, used in conditions of artificial
light, this gives a somewhat higher power output than is
actually produced by the cells. There is considerable debate
as to how the comparison of consumer and power applications
products should be presented. However, since the high added
value of the balance of the consumer product allows a higher
$/W price to be maintained than for the power modules, the
percentage of the revenue would be expected to be higher
than 30%.

Figure 5.17    Small solar panel mounted on car window, for
powering a ventilation fan. (Photograph by
courtesy of Intersolar Ltd.)

Table 5.2       Worldwide PV module shipments

| Application | 1983 | | 1985 | | 1987 | |
|---|---|---|---|---|---|---|
| | MW | Percentage of market | MW | Percentage of market | MW | Percentage of market |
| Consumer products (<10W) | 3.1 | 14.3 | 7.2 | 29.6 | 8.6 | 30.1 |
| Commercial | 8.9* | 41.0 | 11.2 | 45.7 | 18.2 | 63.6 |
| Government** programmes | 9.7* | 44.7 | 6.0 | 24.7 | 1.8 | 6.3 |
| TOTAL | 21.7 | 100 | 24.4 | 100 | 28.6 | 100 |

Data taken from P. D. Maycock, PV News, February issues, 1985, 1986
and 1988.
* These figures were inferred from Maycock, 1985.
** Government programmes include installations for utilities under
third party financing.

It should be noted that the market analysts classify low power as less than 10W for their data. Whilst this is slightly less than has been considered in this chapter, the proportion of low power applications in the 10-40W range has been small compared to those below 10W. In recent years, the number of applications in the 10-40W range has increased and so the total market for low power applications, as defined in this chapter, may be somewhat higher than shown in the table. Whilst the percentage of the total market held by consumer products would be expected to drop as the cost of photovoltaics for large scale applications drops, the market should experience continued growth in real terms.

## 5.7 Future Prospects

Clearly, as shown by the data in the previous section, the low power applications of photovoltaics have experienced rapid market growth in recent years, with the consumer electronics market being the front runner. Although the calculator market appears to be close to saturation, other products are being introduced. The range of products in which photovoltaic power supplies are incorporated is constantly expanding and the consumer products market should grow strongly for some time to come. Large expansion can be forecast in such areas as hand tools, for the kitchen, garden and workshop, incorporating photovoltaics for recharging batteries, and allowing freedom from mains cables either to the hand tool or the recharging unit.

It is not always easy for the scientist or engineer to predict which products will be commercially successful, since much depends on the financial climate, customer preference etc. The photovoltaic consumer products market has benefited from the introduction of marketing experts and their success can be measured in terms of the familiarity of the customer with those products which are widely available. The low power applications market will have come of age when the use of photovoltaic power supplies is standard practice in many leisure products and the customer has come to demand the convenience and reliability of PV power sources in an increasing range of products. Judging by the market growth of the last few years, that day may be approaching rapidly.

## 5.8 References

5.1  Y. Hamakawa and Y. Tawada, Proc. of the 6th E.C. Photovoltaic Solar Energy Conference, London, 1985, [D. Reidel Publishing Co.; 1985].

5.2  K. Takahashi and M. Konagai, Amorphous Silicon Solar Cells, North Oxford Academic, 1986.

5.3  D. E. Carlson, U.S. Patent No. 4,064,521 (1977)

5.4  D. E. Carlson, Proc. of the 3rd E.C. Photovoltaic Solar Energy Conference, Cannes, 1980, [D. Reidel Publishing Co.;1980].

5.5  P. D. Maycock, Proc. of the 6th E.C. Photovoltaic Solar Energy Conference, London, 1985, [D. Reidel Publishing Co.; 1985].

# Index

Applications
  AC power packs, 105
  agricultural, 77, 103
  cathodic protection, 2, 81, 97
  educational, 5, 81, 103
  electric fencing, 78, 104
  grain milling, 78
  grid connected, 13
  instrumentation, 141
  lighting, 36, 63, 104
  low power, 122
    market, 142
  medical
    San Hospital, Mali, 13
  military, 2
  milking machinery, 78
  monitoring, 83
  navigation lights, 2, 83
  professional, 91
  refrigeration, agricultural, 78
  refrigeration, vaccine, 9, 36, 50, 102
  replacement power supplies, 133
  rural electrification, 9, 66
  space, 1
  telecommunications, 2, 34, 80, 99
  television, 5, 103
  terrestrial, 1
  ventilation, 104
  warning lights, 81
  water aeration, 104
  water pumping, 5, 34, 40, 103
  water treatment, 9, 79, 102
Array, 94
  shadowing, 95
  structure, 95

Batteries
  lead acid, 92
  lead calcium, 92
  low antimony, 92
  nickel cadmium, 92
  PV, properties required, 94
  reliability, 84
  sealed, maintenance free, 94
Battery storage, 92

Cathodic protection, 81, 97

Consumer products, 128, 133
  array sizing, 129
  design, 131
  economic feasibility, 131
  introduction, 132
  lifetime, 132
  market, 142
  marketing, 132
  with storage, 132
  without storage, 129
Control units, 92, 95
  solid state, 96
Cooking, 27
  improved stoves, 28
  solar, 29
Costs,
  balance of systems, 19
  batteries, 97
  CCVT, 109
  control units, 97
  grid connected PV systems, 13
  kerosene refrigerators, 60
  photovoltaic modules, 15
  projected
    installation, grid connected, 19
    PV modules, 18
    PV systems, 19
  PV modules, 97
    crystalline silicon, 15
    projected, 18
    thin film silicon, 18
  PV refrigerators, 60
  PV systems, 3, 107
  PV–diesel, 110
  rural electrification, 12, 73, 77
  satellite power supplies, 1
  TEG, 109
  telecommunications systems, 99
  terrestrial systems, 2
  water pumping, 48
Crop spraying, 103

DC–DC conversion, 97
Deforestation, 28
Development
  experience with PV systems, 83
  institutional support, 50, 60, 66, 77, 85

Development (*contd*)
  photovoltaic systems for, 39
  role of energy, 39
  rural electrification, 67
  rural electrication boards, 75
  rural energy areas, 30
Diffusion of photovoltaic systems, 26, 30

Economics
  agricultural systems, 79
  consumer products, 131
  grid connected PV systems, 13
  lighting, 65
  photovoltaic systems, 23, 85
  PV refrigeration, 60
  PV systems, 36, 96
  PV systems for remote homes, 12
  rural electrification, 73
  telecommunications, 81
  telecommunications systems, 97
Education, 103
Electric fencing, 78, 104
Enduse analysis, 27
Energy planning, 26
  appraisal of projects, 32
Energy policy
  Third World, 26

Field testing
  water pumps, solar, 5
Finance
  PV systems
    developing countries, 24
Fota Island, milking system, 78

Giglio Island, cold store, 78

Hybrid systems, 105, 107
  Milton Keynes system, 114
  PV–CCVT, 107, 109
  PV–diesel, 107, 109, 110
  PV–mains, 107, 109
  PV–multielement, 114
  PV–TEG, 107, 109
  PV–wind, 107, 114

Immunisation, 50
Import of technology, 23
Importation of PV modules, 66
Insolation monitoring, 95

Institutional development, 23
Instrumentation, powering by PV, 141
Irrigation
  sprinkler system, 103

Jumuna River, PV warning lights
    system, 81

Kythnos PV pilot plant, 5

Lighting, 63
  domestic, 104
  field experience, 64
  kerosene lamps, 63
  PV systems, 63
  street, 104
Local manufacture, 23
Low power applications
  future prospects, 144
Low power applications of PV, 122

Mali Aqua Viva project, 45
Market
  consumer products, 15
  low power applications, 142
  telecommunications, PV powered,
    15
Market categories, 2
Market development, 21
  obstacles to progress, 22
  rural electrification, 22
  water pumping, 22
Market, photovoltaic, 13
Market prospects, 19
Maximum power point tracking, 92
Milton Keynes
  control system, 117, 119
  solar array, 117
  solar–wind hybrid system, 91, 114
  system configuration, 116
  wind turbine, 117

National energy balance, African
    countries, 27

Photovoltaic array, 94
Photovoltaic market, 13
Photovoltaic systems, 2
Photovoltaics
  terrestrial applications, 1

Photovoltaics, diffusion, 30
  constraints, 31
  enduses, 34
  mechanisms, 32
  project assessment, 30
  social implications, 34
Plants
  demonstration, 3
  demonstration, EEC, 5
  grid connected
    California, 13
  pilot, EEC, 5
    France, 5
    Kythnos, 5
    rural electrification, 70
    village electrification, 72
Power pack, AC, 105
Power supplies
  closed circuit vapour turbine,
    106
  diesel, 106
  mains electricity, 106
  PV, 106
  reliability, 107
  thermoelectric, 106
  wind turbine, 106
Products
  AC power pack, 105
  battery chargers, 2
  consumer, 2, 122, 128, 133
    calculator, 129, 142
    car ventilation, 138
    clock, 132
    market, 142
    with storage, 132
    without storage, 129
  garden lights, 15
  portable crop sprayer, 103
  power supply, 122
  replacement power supplies, 133
    automotive, 138
    leisure, 137
    marine, 138
    military, 137
    portable, 138
  ventilation fans, 104

R&D budget
  European Community, 1
  Japan, 1

US, 1
world, 1
Refrigeration
  benefits of PV, 51
  bottled gas powered, 51
  field experience, 56
  kerosene powered, 51
  PV powered, 9, 50, 102
Regulatory aspects, 24
Reliability
  education systems, 5
  grain milling systems, 78
  lighting systems, 64
  milking systems, 78
  PV refrigerators, 58
  PV systems in developing countries, 84
  PV water pumping systems, 45
  water pumps, 47
Renewable energies
  political commitment, 13
Replacement power supplies, 133
  advantages of PV, 137
  examples, 137
  markets, 137
Rural development, 101
Rural electrification, 9, 66
  advantages of distributed system, 11
  advantages of distributed systems, 68
  commercial equipment, 70
  diesel, 67
  field experience, 72
  load estimation, 67
  subsidies, 11, 75
Rural health care, 50

San Hospital, Mali, 13
Satellite communications systems, 99
Satellite power supplies, 1
Social acceptance, PV systems, 48, 60, 66,
    75, 85
Solamax controller, 92
Solar cells
  crystalline silicon, 15, 94, 123
  current–voltage characteristics, 123
  efficiency, definition of, 125
  lifetimes, 128
  operation, 123
  output parameters, 123
  performance under low illumination,
    126

Solar cells (*contd*)
  spectral response, 125
  stability, 126
  thin film silicon, 15, 95, 123, 142
Sprinklers, agricultural, 103
System design, 96
Systems
  advantages of photovoltaics,
      39
  agricultural, 77
  agricultural sprinkler, 103
  cathodic protection, 81
  consideration of load, 128
  design requirements, 105
  educational, 103
  electric fencing, 104
  grid connected, 13
  hybrid, 91, 105
  hybrid, comparison of candidates,
      107
  lighting, 63, 104
  photovoltaic, 2, 84, 91
  professional, 2, 91
  PV–diesel, 110
  PV–wind, 114
  PV–wind, Milton Keynes, 114
  refrigeration, 102
  refrigeration, PV, 51
  refrigeration, solar powered, 51
  rural electrification, 66, 70, 72
  stand alone, 3
  telecommunications, 80
  water aeration, 104
  water pumping, 37, 103
  water pumping, PV, 45
  water treatment, 102

Tangaye grain mill, 78
Telecommunications, 80, 92, 99
Telecommunications, advances in
      technology, 101
Television
  PV powered, 103

Third World energy sources
  biomass, 27, 29
  charcoal, 28
  coal, 27
  fuelwood, 28
  oil, 26
  renewables (NARSE), 27, 29
Tracking systems, 95
Training, 24

Ventilation, PV powered fans, 104

Water aerator, 104
Water pumping
  drinking water, 5, 40, 103
  economics, 5, 48
  economics, irrigation, 48
  field experience, 45
  irrigation, 3, 5, 40, 43, 103
  Mali Aqua Viva project, 45
  social acceptance of PV, 48
  techniques, 40
Water pumps
  borehole, PV, 43
  centrifugal, 47
  diesel, 40
  electric, 40
  hand, 40
  positive displacement, 47
  PV, 5, 40
  solar powered, 40
  wind, 48
Water sprinkler, 103
Water supply, 43
Water treatment, 79
  filtration, 79
  UV, 80, 102
Wind power applications
  grid connected, 13
World Health Organisation
  approved PV refrigerator suppliers, 56
  refrigeration projects, 51
  specification for refrigerators, 52